# Urban Ants
# of North America
# and Europe

# URBAN ANTS
## of North America and Europe

*Identification, Biology, and Management*

**JOHN KLOTZ**
*University of California, Riverside*

**LAUREL HANSEN**
*Spokane Falls Community College*

**REINER POSPISCHIL**
*Bayer CropScience Aktiengesellschaft*

**MICHAEL RUST**
*University of California, Riverside*

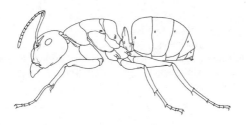

**Comstock Publishing Associates** A DIVISION OF

Cornell University Press  *Ithaca and London*

First published 2008 by Cornell University Press

First printing, Cornell Paperbacks, 2008

Printed in the United States of America

*Library of Congress Cataloging-in-Publication Data*
Urban ants of North America and Europe : identification, biology, and management / John Klotz ... [et al.].
   p. cm.
  Includes bibliographical references and index.
  ISBN 978-0-8014-7473-6 (pbk. : alk. paper)
   1. Ants—North America.  2. Ants—Europe.  3. Urban pests—North America.
4. Urban pests—Europe.  I. Klotz, John H., 1946–  II. Title.
  QL568.F7U725 2008
  595.79′6—dc22
2008017829

Cornell University Press strives to use environmentally responsible suppliers and materials to the fullest extent possible in the publishing of its books. Such materials include vegetable-based, low-VOC inks and acid-free papers that are recycled, totally chlorine-free, or partly composed of nonwood fibers. For further information, visit our website at www.cornellpress.cornell.edu.

Paperback printing   10 9 8 7 6 5 4 3 2 1

# Contents

BIOLOGICAL CONTROL    157
INTEGRATED PEST MANAGEMENT (IPM)    159
CONCLUSION    161

# Preface

In recent years the structural pest control industry has consistently ranked ants as the number one pest, surpassing termites, cockroaches, and rodents; most homeowners regard them as more troublesome than cockroaches. Ants represent a diverse and expansive group of insects, especially in the tropics. Approximately 11,800 species have been described (Moreau et al. 2006), and perhaps as many as 20,000 extant species exist on the Earth (Hölldobler and Wilson 1990). Most species are confined to certain regions, although some "tramp ants" have a worldwide distribution. Some species, such as the eastern black carpenter ant, pharaoh ant, and odorous house ant, may nest within structures; others, such as the Argentine ant, nest outdoors. The fire ants and harvester ants tend to occupy landscaped areas surrounding homes, parks, and recreation areas.

In the United States alone, ant control generates an estimated $1.7 billion annually for pest management professionals (PMPs). If the red imported fire ant were to spread throughout California, the total financial impact on households would probably be at least $342 million (Jetter et al. 2002). The threat is so real that residents of Orange County, California, recently approved an increase in property assessments in part to control red imported fire ants.

Effective treatment of ant infestations requires correct identification, something that challenges even the best extension specialists and PMPs. We hope to alleviate that problem by providing comprehensive keys to the species likely to be encountered in urban environments. The illustrations focus on morphological features that permit rapid identification. The species accounts provide information on behavioral and ecological habits that will further assist in identification. Numerous common names are in use for some of these species, but we have adopted those approved by the Entomological Society of America (ESA) (http://www.ent.soc.org/Pubs/common_Names/index.htm). Common names frequently used in the literature but not formally approved are inside quotation marks, such as *Myrmica rubra,* the "European fire ant." These so-called common names often have a regional origin and then later gain universal acceptance.

Our focus is on the pest ants of urban environments in the United States, Canada, and Europe. Some of these species pose a serious medical threat to humans and their domesticated pets; others may do serious damage to property and landscapes. Not all species pose a significant problem, although their mere presence may be sufficient to result in control efforts. It is our hope that this book will serve as a reference not only for PMPs, urban extension specialists, homeowners, and regulatory agencies, but for those interested in insects associated with humans and urban environments.

Practical information about the control of urban pest ants is limited to a few species such as Argentine ants, carpenter ants, fire ants, and pharaoh ants. Comprehensive integrated pest management programs exist only for red imported fire ants. We hope that this book will stimulate researchers and PMPs to study the other important pest species. Consequently, we do not provide a "cookbook" approach to control. We hope that PMPs and homeowners will use it as a reference and a starting point to design their own control programs.

Basic research on ants has provided the foundation on which management strategies are built. E. O. Wilson and Bert Hölldobler are the icons of this realm. The efforts of the Agricultural Research Service of the U.S. Department of Agriculture (USDA) to control the red imported fire ant have served as a model for applied ant research for the last fifty years (review Williams et al. 2001). Stoy Hedges has provided the most comprehensive approach to ant control to date for pest management professionals. His *Field Guide for the Management of Structure-Infesting Ants* is a landmark publication in this regard. Stoy has dedicated his career to educating the pest control industry on good management practices not only for ants but for many other structural pests as well. We hope that our efforts follow in his path. It is in this spirit that we dedicate this book to him for his noteworthy efforts and contributions to this challenging, rapidly changing, and ever-evolving field.

In dedication to Stoy Hedges

# Acknowledgments

Addressing the broad range of topics necessary in a book on urban ants and their management required the unique strengths of each author. Several others who generously provided their expertise made significant contributions as well. Stephen A. Klotz, M.D., reviewed most of the chapters and provided invaluable editorial advice. Jacob Pinnas, M.D., offered guidance on the medical aspects of stinging ants, and Justin Schmidt contributed a critical review of that chapter. Greg Vogel and Mary Ann Stanley did a superb job editing the introduction. Lloyd Davis Jr. and Michael Martinez provided invaluable taxonomic advice. Janet Reynolds illustrated figure 1.3 (anatomy of an ant) and figures 2.14–2.24 (all on Camponotus). Sarah Alderete contributed the remaining illustrations in the taxonomic keys. Sharon Carroll compiled and edited the illustrations and photographs, and prepared the index. Most important of all were the professional services that Cornell University Press provided from beginning to end in the production of this book. Unless otherwise indicated, all photographs are by the authors.

Keys are adapted from a number of sources, including (in alphabetical order) Bolton 1994; Bolton and Collingwood 1975; Collingwood 1979; Coovert 2005; Creighton 1950; Czechowski et al. 2002; Gregg 1963; Hansen and Klotz 2005; Hedges 1998; Hölldobler and Wilson 1990; Seifert 1988, 1992, 1996; Smith 1965; Snelling 1988; Snelling and George 1979; Trager 1984; Wheeler and Wheeler 1963, 1973, 1986; Wilson 2003; Wilson and Taylor 1967; plus Internet sources: Ant Web; Ants of Africa; Smithsonian Institution/Discover Life. Taxonomic classification follows Bolton et al. 2006.

# Urban Ants
# of North America
# and Europe

**Plate 1**

a. *Camponotus floridanus* major worker

b. *Camponotus ligniperdus* major worker

c. *Formica rufa* workers

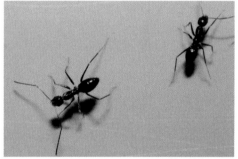

d. *Paratrechina longicornis* workers (Photograph by Sanford Porter, USDA, ARS, CMAVE)

e. *Prenolepis imparis* worker

f. *Lasius niger* queen with workers and larvae

## Plate 2

a. *Anoplolepis gracilipes* workers (Photograph by Neil Reimer, Hawaii State Department of Agriculture)

b. *Plagiolepis* sp. queen with workers and pupa

c. *Linepithema humile* workers with citrus mealybugs (Photograph by Dong-Hwan Choe, University of California, Riverside)

d. *Tapinoma sessile* queen with workers

e. *Tapinoma melanocephalum* queens with workers

f. *Technomyrmex albipes* workers

**Plate 3**

a. *Liometopum occidentale* workers

b. *Monomorium pharaonis* queen with workers and brood

c. *Monomorium destructor* queen with workers

d. *Pheidole megacephala* major and minor workers

e. *Tetramorium caespitum* worker

f. *Wasmannia auropunctata* worker and larva (Photograph by Sanford Porter, USDA, ARS, CMAVE)

# Plate 4

a. *Pogonomyrmex rugosus* workers
(Photograph by Alexander Yelich)

b. *Solenopsis invicta* male with workers

c. *Crematogaster scutellaris* workers

d. *Myrmica rubra* workers with larvae

e. *Hypoponera punctatissima* winged female
(Photograph by Gary Alpert, permission
granted from *Annals of Allergy, Asthma and
Immunology* 2005 [95: 418–425])

f. *Pachycondyla chinensis*

# CHAPTER ONE

# Biology and Ecology of Pest Ants

## Origins and Social Behavior

Since their origin in the Mesozoic, ants have undergone an adaptive radiation of impressive proportions generating a diversity of species that have filled myriad ecological niches (Wilson 1971). The fossil evidence indicates that ants arose from a vespoid wasp ancestor. *Sphecomyrma freyi* Wilson & Brown, an eighty-million-year-old primitive ant discovered in 1967 in New Jersey amber, had an antlike petiole, wasplike mandibles, and antennae intermediate between those of modern-day wasps and ants (Hölldobler and Wilson 1990). Unlike wasps, however, all ants are social and live in colonies.

Several factors may have predisposed ants toward sociality. Their haplodiploid mode of sex determination, whereby males are derived from haploid eggs (one set of chromosomes) and females from diploid eggs (two sets of chromosomes), has been conducive to the development of a sterile worker caste consisting of females that forgo reproduction to raise more closely related sisters. Their multipurpose jaws enable ants to defend the colony, subdue and carry prey, transport brood, and excavate nests. Their relatively large brain enables ants to leave their nest, go out into a complex environment—sometimes several hundred feet away—in search of food, and then return with it to feed the colony.

An individual worker leaves the nest on a specific trail that leads to a particular foraging site, where it feeds on a certain food. Thus, the forager demonstrates trail, site, and resource fidelity while using chemical and sometimes physical trails to conserve foraging time. Ants follow natural edges as well as structural elements, further reducing their foraging time and probably facilitating navigation (Klotz et al. 2000), and they use visual cues for orientation, including the sun, patterns of polarized light, and landmarks. Some species can even use the geomagnetic field of the Earth as an orientation cue (Jander and Jander 1998). Central-place foraging theory assumes that foragers become more selective of food the farther they must travel to obtain it (Hölldobler and Wilson 1990).

## Ecological and Economic Importance

Ants play critical roles in most ecosystems. For example, they are primary predators of forest-defoliating insects, especially during outbreaks, when these pests can make up more than 90% of an ant's diet (Petal 1978). Harvester ants are among the chief granivores in deserts (Davidson et al. 1980). Ants also aerate soil, decompose detritus, recycle nutrients, and pollinate plants. In Europe, wood ants of the *Formica rufa* group, particularly *F. rufa* and *F. polyctena,* are cultured for biological control in forests and are strictly protected. Notwithstanding their valuable contributions, ants can also be detrimental. Their most significant agricultural economic impact results from their tending of aphids, mealybugs, and scales and interfering with parasitoids in biological control programs. As stinging Hymenoptera they are not as important as wasps and bees, but in certain geographic regions more than 50% of the human population has been stung by imported fire ants (deShazo et al. 1990). Some species of fire ants and other field ants are important pests in parks, recreational areas, and golf courses. Other ants are household pests, excavating in wood frameworks and raiding food stores.

In the United States, ant control tops the percentage of money spent on structural pest control (Whitford 2006), generating an estimated $1.7 billion for pest management professionals (PMPs) each year (Curl 2005). Forty-one species of ants are considered household pests in the United States (Hedges 1998), and ten or so species have attained major economic pest status in Europe (Weidner and Sellenschlo 2003; Pospischil 2005). Nevertheless, only a tiny fraction of the twenty thousand or so extant species (Hölldobler and Wilson 1990) have become urban pests.

The relative importance of each pest ant species varies according to geographic location. PMPs in the Pacific Northwest, for example, most frequently deal with carpenter ants (Hansen and Akre 1985), which in some cases account for about 35% of the pest control business (T. Whitworth, pers. comm., 2003). Argentine ants are the most common ant pests in California—and the most difficult to control (Knight and Rust 1990a). A large pest control firm in San Diego with approximately thirty-five thousand general pest accounts reported that 85% of its service calls for ants were to control Argentine ants (Field et al. 2007).

## Native and Exotic Pest Ants

Urban pest ants are a mixture of native and exotic species. Of the 281 ant species in California, 22 are exotic (Ward 2005). Carpenter ants (*Camponotus* species) are probably most familiar to homeowners, although other household pests such as odorous house ants (*Tapinoma sessile*) and velvety tree ants (*Liometopum occidentale*) are also native to California. By far the most economically important pest ants in California are exotic invaders: the Argentine ant (*Linepithema humile*) and the red imported fire ant (*Solenopsis invicta*). They are significant pests not only in urban and agricultural environments but also in natural areas, where they cause ecological damage by displacing other native species of ants (Vinson 1997; Vega and Rust 2001).

Until recently, the pharaoh ant was the only exotic household pest ant reported in Europe. However, the number of introduced species has been increasing (Cornwell 1978). In Switzerland, for example, seven exotic species of household ants other than pharaoh ants were reported in 2005 (Umwelt- und Gesundheitsschutz Zuerich 2004).

Worldwide, 147 species of ants have been recorded living in nonnative habitats (McGlynn 1999). Exotic species are thus not a new problem, but their frequency is increasing because of globalization and urbanization. Increased trade and travel along with the disruption of natural ecosystems have prepared the way for introduced species that thrive in disturbed habitats. Once a new species gains a foothold, it may proliferate to varying degrees. The species considered invasive spread aggressively into both disturbed and undisturbed habitats, outcompeting and displacing native ant species (McGlynn 1999). Another group called tramp ants are tied to human activities; as their name implies, they are easily transferred from one location to another (Hölldobler and Wilson 1990). Some introduced species exhibit characteristics of both groups (McGlynn 1999).

In addition to unintentional transfer, there is a burgeoning "pet trade" market in live insects, including ants. In Europe, for example, hobbyists interested in rearing ants in a terrarium or—in the case of certain species of *Pheidole*—

as food for animals such as exotic arrow poison frogs can order small colonies from Internet sources. The ants, particularly the polygynous species, may escape and become established in new areas (Buschinger 2004). Culturing ants at home is also becoming popular in the United States, although it is illegal to trade ant queens.

## Life History Traits

Most invasive and tramp species share several life history characteristics that contribute to their success in colonizing new environments (Passera 1994). They are unicolonial, meaning that they have open societies that lack intraspecific aggression. Thus, there is no clear delineation between a colony and a population, and individual ants readily move from one nest to another. The colonies are polygynous as well and reproduce by fission or budding. When a colony buds, one or more of the many queens in the parent colony leave the nest along with some workers and brood to form a new colony.

Colony budding eventually creates a network of interrelated nests that form a cooperative unit, which may extend over an entire habitat. The flow of food in these supercolonies is decentralized; that is, its movement depends on the needs of the individual colonies, a behavior known as dispersed central-place foraging (McIver 1991; Holway and Case 2000). Instead of expending valuable energy on colony defense, these large cooperative units channel their energy into foraging and colony growth. Supercolonies can outcompete native species for limited resources, and under certain conditions may saturate the environment. For instance, a researcher in Louisiana reported trapping 1,307,222 Argentine ant queens and 4,352 l (1,150 gal) of workers and brood over a 1-year period in a 7.7-ha (19 ac) citrus grove (Horton 1918).

A few of the native urban pest ants possess these characteristics as well, but on a more limited scale; as, for example, colonies of *T. sessile,* which lack intraspecific aggression within a small area or single habitat. More often, native species have multicolonial, closed societies that are territorial, monogynous, and achieve colony multiplication by sending out winged (alate) males and females. These "reproductives" mate, and each newly fertilized queen forms a new colony.

Mating flights, usually triggered by weather cues, take place in a synchronized fashion over a large geographic area. Winged males and females fly to a congregation area, sometimes marked by a conspicuous landmark such as a hilltop or particular tree or bush, and either mate in the air (e.g., carpenter ants) or on the ground (e.g., many species of harvester ants). After mating, the male dies and the inseminated female leaves in search of a nest site. When she finds a suitable location, she removes her wings, excavates a nest chamber, seals the entrance to the chamber, and begins laying eggs. Only a few of the eggs will reach maturity; most will serve as food for the queen or the faster-developing larvae. The queen also obtains nutrition from her fat reserves and now-non-functional wing muscles, which are broken down inside the body and resorbed.

The developmental stages of ants include eggs, larvae, pupae, and adults. The time required to complete each stage varies with species and environmental conditions. Eggs are small, and those of each species have a characteristic shape. The white, legless larvae develop through three to five instars (stages), depending on the species. Each larva has a head capsule with mandibles for eating. Mature larvae of some species spin a cocoon composed of fine silk that hardens into a brownish or whitish papery material inside which they pupate.

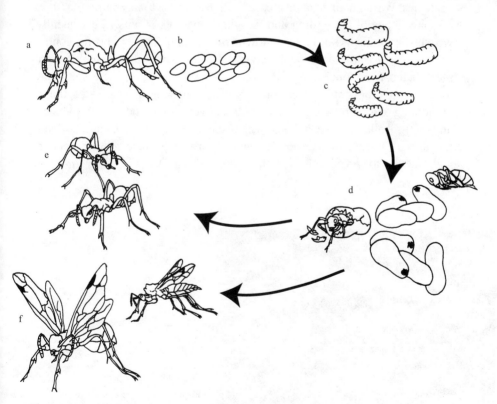

**Figure 1.1.** Life cycle of ants. (a) Queen (b) Eggs (c) Larvae (d) Pupae (e) Workers (f) Winged reproductives

The first workers that emerge care for the subsequent brood as it develops. The nest expands and foraging begins. The queen's sole duty now is egg-laying. After a few seasons the colony will reach considerable strength, and a new group of sexual alates will be produced and will leave the nest to mate (Fig. 1.1).

The red imported fire ant is unique in having evolved both of the social forms found in ants: a nonterritorial, polygyne form that has colonies with many queens and can found new colonies by budding; and a territorial, monogyne form that has one queen per colony and sends out winged reproductives on mating flights.

**Implications for Control**

Understanding colony dynamics is critical for developing pest management strategies. We maintain that the most effective and least toxic control measures take advantage of the nesting and foraging behaviors of ants. Consequently, we emphasize the use of baits or slow-acting, nonrepellent sprays for control because they exploit the social behavior of ants and are safer for the environment than conventional sprays. Baits and slow-acting nonrepellent sprays are ideally suited for central-place foragers such as ants, because they capitalize on homing and recruitment behaviors to deliver an insecticide into the nest, and on trophallaxis (a behavior in which ants feed one another) and horizontal transfer (movement of an active ingredient from one ant to another by physical contact), to spread it through the colony (Fig. 1.2).

**Figure 1.2.** Trophallaxis between two carpenter ant (*Camponotus floridanus*) majors

## Biology and Taxonomy

Their social organization and food-gathering habits make ants among the most familiar and easily recognized insect groups. All ants, being eusocial, exhibit cooperative brood care, reproductive division of labor (fertile queens lay eggs, and sterile workers do the colony's "housekeeping"), and overlapping generations (parent and offspring) that contribute to colony labor (Wilson 1971).

Each adult belongs to a specific caste, the primary ones being queens, males, and workers. Queens are the largest members of the colony. Males are typically intermediate in size between queens and workers and have a pointed abdomen with terminal genitalia. In most cases, both queens and males are alate reproductives before mating. After mating, the queens remove their wings (dealate) and the males die. Winged females and males have ocelli; the workers of some species have ocelli as well.

The wingless workers have elbowed antennae and a narrow constriction called the petiole at the beginning of the gaster. The petiole is composed of one or two segments that bear an upright projection called the node (Fig. 1.3). Identification keys are based primarily on the worker caste.

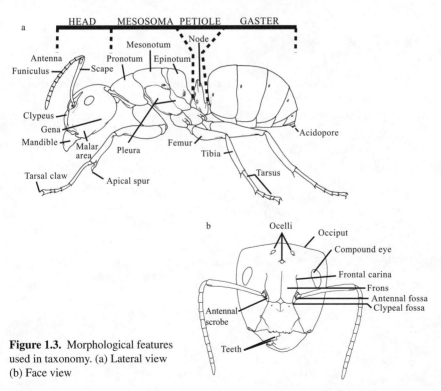

**Figure 1.3.** Morphological features used in taxonomy. (a) Lateral view (b) Face view

Homeowners often confuse the alates of ants with those of termites. Three characteristics easily differentiate the two groups: (1) the front and hind wings of ants differ in size, while those of termites are similar in size; (2) the antennae of female ants are elbowed (those of males are threadlike or, rarely, elbowed), and those of termites are straight and beadlike; and (3) the abdomen of ants is constricted at the base of the thorax while that of termites is broadly joined to the thorax.

The order in which the ants are presented in this book is based on taxonomy and has no relation to their significance as urban pests. Ant taxonomy is currently undergoing rapid and revolutionary changes. Not long ago, when William Creighton's *Ants of North America* (1950) was the most widely consulted identification key, ants were divided into seven subfamilies. Hölldobler and Wilson's *The Ants* (1990) recognizes eleven subfamilies. Bolton's *Synopsis and Classification of Formicidae* (2003) lists twenty-one subfamilies (subsequently reduced to twenty subfamilies [Moreau et al. 2006]) and 288 genera. Most of the urban pest ants belong to the subfamilies Formicinae, Dolichoderinae, and Myrmicinae, which provide the major divisions of this book.

### Key to Subfamilies

1 Abdominal pedicel composed of 2 segments: petiole and postpetiole (Fig. 1.4a); sting usually present . . . . . . . . . . . . . . . . . . . . . . . . . . . . . . . . . 2

  Abdominal pedicel composed of 1 segment (Fig. 1.4b) . . . . . . . . . . . . 4

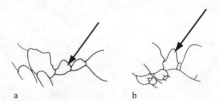

a                  b

**Figure 1.4.** Profile of abdominal pedicel. (a) *Pheidole* spp. (arrow: two-segmented pedicel) (b) *Camponotus* spp. (arrow: one-segmented pedicel)

2(1) Eyes lacking or extremely small, ocellus-like (Fig. 1.5a) . . . . . . . . . . . . . . . . . . . . . . . . . . . . . . . . . . . . . . . . . . . . . . . . . . . Ecitoninae, Chapter 7

     Large compound eyes (Fig. 1.5b) . . . . . . . . . . . . . . . . . . . . . . . . . . . . . 3

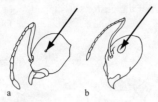

**Figure 1.5.** Profile of head. (a) *Neivamyrmex opacithorax* (arrow: small compound eye) (b) *Camponotus* spp. (arrow: large compound eye)

3(2) Eyes unusually large, covering most of sides of head; first mesosomal segment articulates freely with second mesosomal segment (Fig. 1.6a) . . . . . . . . . . . . . . . . . . . . . . . . . . . . . . . . . . . . . . . . . . . Pseudomyrmecinae, Chapter 6

Eyes not unusually large; first and second mesosomal segments fused (Fig. 1.6b) . . . . . . . . . . . . . . . . . . . . . . . . . . . . . . . . . Myrmicinae, Chapter 4

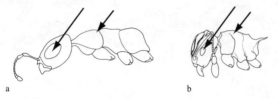

**Figure 1.6.** Profile of head and mesosoma. (a) *Pseudomyrmex* spp. (arrows: large compound eye, articulated first mesosomal segment) (b) *Wasmannia auropunctata* (arrows: compound eye not enlarged, mesosomal segments fused)

4(1) Node broad at the apex, rectangular (Fig. 1.7) . . . . Ponerinae, Chapter 5

Node not broad at the apex, flattened or pointed (Fig. 1.4b) . . . . . . . . . 5

**Figure 1.7.** Profile of pedicel and gaster of *Hypoponera punctatissima* (arrow: broad apex of node)

5(4) Tip of gaster with circular acidopore usually surrounded by a fringe of hairs (Fig. 1.8a) . . . . . . . . . . . . . . . . . . . . . . . . . . . . . . . Formicinae, Chapter 2

Opening at the end of the gaster slit-shaped, not surrounded by hairs (Fig. 1.8b) ............................... Dolichoderinae, Chapter 3

**Figure 1.8.** Profile and ventral views of tip of gaster. (a) *Camponotus* spp. (arrow: acidopore) (b) *Liometopum* spp. (arrow: transverse, slitlike orifice)

# CHAPTER TWO

# Formicinae

## Subfamily Characteristics

Members of this cosmopolitan subfamily have a one-segmented petiole in the form of a vertical scale. The anus is terminal and is usually surrounded by a circular fringe of hairs. These ants do not sting but can spray venom that contains formic acid.

### SCIENTIFIC AND COMMON NAMES

*Anoplolepis gracilipes* (F. Smith, 1857): Long-legged ant
*Brachymyrmex:* Rover ants
    *B. patagonicus* Mayr, 1868
    *B. depilis* Emery, 1893
    *B. obscurior* Forel, 1893
*Camponotus:* Carpenter ants
  Subgenus *Camponotus*
    *C. americanus* Mayr, 1862
    *C. chromaiodes* Bolton, 1995
    *C. herculeanus* (Linnaeus, 1758)
    *C. laevigatus* (F. Smith, 1858)
    *C. ligniperdus* (Latreille, 1802)
    *C. modoc* W.M. Wheeler, 1910: "Western black carpenter ant"
    *C. novaeboracensis* (Fitch, 1855)
    *C. pennsylvanicus* (DeGeer, 1773): Black carpenter ant
    *C. vagus* (Scopoli, 1763)

Subgenus *Colobopsis*
    *C. truncatus* (Spinola, 1808)
Subgenus *Myrmentoma*
    *C. caryae* (Fitch, 1855)
    *C. clarithorax* Creighton, 1950
    *C. decipiens* Emery, 1893
    *C. discolor* (Buckley, 1866)
    *C. essigi* M.R. Smith, 1923
    *C. fallax* (Nylander, 1856)
    *C. hyatti* Emery, 1893
    *C. nearcticus* Emery, 1893
    *C. piceus* (Leach, 1825)
    *C. sayi* Emery, 1893
    *C. subbarbatus* Emery, 1893
Subgenus *Myrmobrachys*
    *C. planatus* Roger, 1863
Subgenus *Myrmothrix*
    *C. floridanus* (Buckley, 1866): Florida carpenter ant
Subgenus *Taenamyrmex*
    *C. acutirostris* W.M. Wheeler, 1910
    *C. castaneus* (Latreille, 1802)
    *C. semitestaceus* Snelling, 1970
    *C. tortuganus* Emery, 1895
    *C. variegatus* (F. Smith, 1858)
    *C. vicinus* Mayr, 1870
*Formica:* Thatching and field ants
 *exsecta* group
    *F. exsectoides* Forel, 1886
 *fusca* group
    *F. argentea* W.M. Wheeler, 1912
    *F. francoeuri* Bolton, 1995
    *F. fusca* Linnaeus, 1758
    *F. neorufibarbis* Emery, 1893
    *F. subsericea* Say, 1836
 *microgyna* group
    *F. densiventris* Viereck, 1903
 *neogagates* group
    *F. perpilosa* W.M. Wheeler, 1902
 *pallidefulva* group
    *F. nitidiventris* Emery, 1893
 *rufa* group: Wood ants
    *F. integroides* W.M. Wheeler, 1913
    *F. obscuripes* Forel, 1886
    *F. obscuriventris* Mayr, 1870

*F. oreas* W.M. Wheeler, 1903
*F. planipilis* Creighton, 1940
*F. polyctena* Foerster, 1850
*F. pratensis* Retzius, 1783
*F. propinqua* Creighton, 1940
*F. ravida* Creighton, 1940
*F. rufa* Linnaeus, 1761
*F. subnitens* Creighton, 1940
*F. truncorum* Fabricius, 1804
*sanguinea* group
    *F. aserva* Forel, 1901
    *F. subintegra* W.M. Wheeler, 1908
*Lasius:*
Subgenus *Acanthomyops:* Citronella ants
    *L. claviger* (Roger, 1862): "Small yellow ant"
    *L. interjectus* Mayr, 1866: "Larger yellow ant"
    *L. latipes* (Walsh, 1863)
    *L. murphyi* Forel, 1901
Subgenus *Cautolasius*
    *L. flavus* (Fabricius, 1782)
Subgenus *Chthonolasius*
    *L. subumbratus* Viereck, 1903
    *L. umbratus* (Nylander, 1846)
Subgenus *Dendrolasius*
    *L. fuliginosus* (Latreille, 1798)
Subgenus *Lasius*
    *L. alienus* (Foerster, 1850): Cornfield ant
    *L. brunneus* (Latreille, 1798)
    *L. emarginatus* (Olivier, 1792)
    *L. neoniger* Emery, 1893
    *L. niger* (Linnaeus, 1758): "Black garden ant"
    *L. pallitarsus* (Provancher, 1881)
    *L. paralienus* Seifert, 1992
    *L. platythorax* Seifert, 1991
    *L. psammophilus* Seifert, 1992
    *L. neglectus* Van Loon, Boomsma & Andrasfalvy 1990
*Paratrechina*
    *P. bourbonica* (Forel, 1886)
    *P. longicornis* (Latreille, 1802): Crazy ant
    *P. pubens* (Forel, 1893)
    *P. vividula* (Nylander, 1846)
*Plagiolepis* spp.
*Prenolepis imparis* (Say, 1836): "Small or false honey ant" or "winter ant"

## Key to Genera of Formicinae

1  Antenna with 9 segments (Fig. 2.1a) ............ *Brachymymex* spp.
   Antenna with 11 or 12 segments (Fig. 2.1b, c) .................... 2

**Figure 2.1.** Antenna. (a) *Brachymyrmex* spp. (b) *Camponotus* spp. (c) *Plagi-olepis* spp.

2(1)  Antenna with 11 segments (Fig. 2.1c) .............. *Plagiolepis* spp.
     Antenna with 12 segments (Fig. 2.1b) ......................... 3

3(2)  Profile of thoracic dorsum evenly convex (Fig. 2.2a) .... *Camponotus* spp.
     Profile of thoracic dorsum with epinotum distinctly depressed below the
     level of mesonotum (Fig. 2.2b) ................................ 4

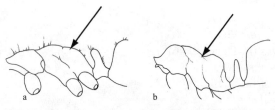

**Figure 2.2.** Mesoma profile. (a) *Camponotus* spp. (arrow: smooth thoracic dorsum) (b) *Lasius* spp. (arrow: uneven thoracic dorsum)

4(3)  Epinotal spiracle a narrow slit (Fig. 2.3a); frontal carina prominent .....
     ............................................... *Formica* spp.
     Epinotal spiracle rounded (Fig. 2.3b); frontal carina feebly developed ...
     ............................................................ 5

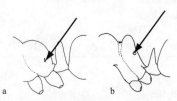

**Figure 2.3.** Profile of epinotum and pedicel. (a) *Formica* spp. (arrow: slitlike epinotal spiracle) (b) *Lasius* spp. (arrow: round epinotal spiracle)

5(4)  Scape surpassing occipital margin by less than one-third its length (Fig.
     2.4a) ......................................... *Lasius* spp.

Scape surpassing occipital margin by at least one-third its length (Fig. 2.4b) ................................................. 6

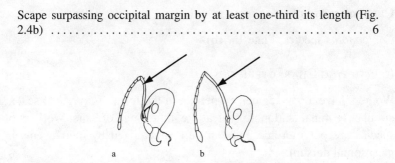

**Figure 2.4.** Profile of head and antenna. (a) *Lasius* spp. (arrow: shorter scape) (b) *Prenolepis imparis* (arrow: longer scape)

6(5) Mesosoma strongly constricted; swollen in front of and behind constriction (Fig. 2.5a); scape and tibiae without erect hairs ................. ........................................... *Prenolepis imparis* Thorax only slightly constricted at mesosoma (Fig. 2.5b); scape and tibiae with erect hairs ........................................... 7

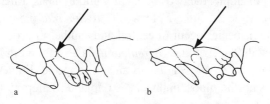

**Figure 2.5.** Profile of mesosoma. (a) *Prenolepis imparis* (arrow: mesosomal constriction) (b) *Paratrechina longicornis* (arrow: mesosoma not constricted)

7(6) Yellow color; dorsum of mesosoma nearly devoid of standing hair; mesonotum in profile weakly concave (Fig. 2.6) ..... *Anoplolepis gracilipes* Dark brown color; dorsum of mesosoma bearing numerous long, erect hairs; mesonotum in profile weakly convex (Fig. 2.5b) .............. ............................................. *Paratrechina* spp.

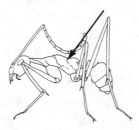

**Figure 2.6.** Profile of *Anoplolepis gracilipes* (arrow: concave mesosoma)

## Carpenter Ants

*Camponotus* species (Plate 1a, 1b)

### IDENTIFYING CHARACTERISTICS

Workers have a one-segmented petiole in the form of a vertical scale, and a circular, terminal acidopore typically with a fringe of hairs. Workers of most species are polymorphic and can be characterized by their evenly convex mesosomal dorsum.

### DISTRIBUTION

*Camponotus* is a huge, hyperdiverse, and cosmopolitan genus rivaled in size only by *Pheidole* (Wilson 2003). Twenty-four species of *Camponotus* are structural or nuisance pests in the United States (Hansen and Klotz 2005). Several species originate in Europe, but only *C. herculeanus* and *C. ligniperdus* have attained pest status (see Table 2.1 for geographic distributions of species). In Sweden, for example, the two are equally important as structural pests (Butovitsch 1976), while in southern Norway *C. ligniperdus* predominates (Birkemoe 2002). Both species occur in central and southern Europe (Seifert 1996; Czechowski et al. 2002), and *C. herculeanus* is also found in North America and Asia. In Europe, *C. herculeanus* is an alpine species limited to mountainous regions; its northernmost boundary is Lapland (Seifert 1996).

### BIOLOGY AND HABITS

Carpenter ants enter buildings chiefly to establish satellite nests. They are called "carpenters" because they create smooth tunnels and galleries in wood (Fig. 2.7). They generally choose wood that is decayed or damaged by other insects, but they also nest in structurally sound wood (Fowler 1986).

Carpenter ant colonies are established after the mating flights of winged male and female reproductives. These nuptial flights usually take place on warm spring days following a cool, rainy period. After mating, the males die and the inseminated females disperse to search for nest sites (Fig. 2.8). The colonies of most species are monogynous and are founded by a single queen. A queen of *C. modoc,* for example, often lays her first ten–twenty-five eggs in a small cavity in a dead or live tree. In 2 to 3 weeks the eggs hatch into larvae, which the queen provides with nourishment derived from her fat bodies and metabolized wing muscles. At the end of the fourth and final instar, these larvae pupate, later to emerge as minor workers that immediately begin foraging, excavating, and brood rearing for the colony.

**Table 2.1.** Generalized distribution pattern of *Camponotus* species in North America and Europe

| Subgenus | Species | Approximate geographic distribution |
|---|---|---|
| *Camponotus* | *americanus* | Eastern and central U.S. and southeastern Canada |
| | *chromaiodes* | Eastern and central U.S. and southeastern Canada |
| | *herculeanus* | Northeastern and western U.S., Canada, and into AK; Northern and eastern Europe, north beyond Arctic Circle |
| | *laevigatus* | Western North America |
| | *ligniperdus* | Europe north beyond Arctic Circle, south to northern Spain, Italy, and Greece |
| | *modoc* | Western U.S. and southwestern Canada |
| | *novaeboracensis* | Northern U.S. and southern Canada, ND south into NM; recorded in central TX |
| | *pennsylvanicus* | Eastern and central U.S. and southeastern and south-central Canada |
| | *vagus* | Europe, north to southern Finland |
| *Colobopsis* | *truncatus* | Northern Europe |
| *Myrmentoma* | *caryae* | NY west to IA and KS, and south to FL |
| | *clarithorax* | CA, OR, and northern part of Baja California, Mexico |
| | *decipiens* | GA and FL west to TX and eastern Mexico, and north to ND |
| | *discolor* | Northeastern Mexico and TX through the southeastern states and north to ND |
| | *essigi* | Northwestern Mexico to southern Canada |
| | *fallax* | Europe |
| | *hyatti* | Southern CA through OR, east to the Rocky Mt. states, and into northwestern Mexico |
| | *nearcticus* | Throughout southern Canada and eastern and northern parts of the U.S., extending south through the central states |
| | *piceus* | Southern Europe |
| | *sayi* | Southwestern U.S. and Mexico, extending from southern CA north to NV and UT, and east to KS, NE, and TX |
| | *subbarbatus* | Along the Atlantic Coast from New England to SC, GA, and MS, and westward into OH, KY, and TN |
| *Myrmobrachys* | *planatus* | Southern FL, southern TX, and northeastern Mexico |
| *Myrmothrix* | *floridanus* | FL north to NC and west to MS and LA |
| *Tanaemyrmex* | *acutirostris* | NM and TX |
| | *castaneus* | KS and IA east to NY and southern New England, and south to TX and FL |
| | *semitestaceus* | Coastal ranges of CA as far north as San Francisco Bay area |
| | *tortuganus* | Southern FL |
| | *variegatus* | Hawaiian Islands |
| | *vicinus* | Mexico to southern Canada, inland to the Rocky Mts. and east to Alberta, the Dakotas, and KS |

*Sources:* Snelling 1988, 2006; Seifert 1996; Czechowski et al. 2002; Hansen and Klotz 2005.

17

**Figure 2.7.** Sand-papered appearance of damage to wood by carpenter ant (*Camponotus modoc*)

Within 2 years, a population of workers ranging in size from small minors to large majors will be present. Within 3 to 5 years, colonies of *C. pennsylvanicus* start producing alates (Pricer 1908). Mature *C. modoc* colonies may contain more than 50,000 workers, and those of *C. vicinus* more than 100,000 (Akre et al. 1994a). This vast difference in numbers is partly due to the presence of multiple queens in *C. vicinus* colonies; as many as forty-one functional

**Figure 2.8.** Winged female carpenter ant (*Camponotus ligniperdus*)

**Figure 2.9.** Carpenter ant (*Camponotus floridanus*) workers with pupae

queens have been collected from a single colony (Akre et al. 1994a). Most species of carpenter ants, however, are monogynous, and consequently have smaller colonies.

Mature colonies establish satellite colonies nearby whenever a need exists for more territory, more resources, or a drier or warmer nesting site for development of their larvae and pupae. The queen, workers, and small larvae remain in the parent colony, and workers move the larger larvae and pupae into the satellite colonies (Hansen and Akre 1990). Winged reproductives brought in as pupae during the summer months may also overwinter in these satellite colonies (Fig. 2.9). Except during the winter, workers travel between the various satellites of a colony on well-defined trails. The distance between parent and satellite nests varies, but in *C. modoc* colonies has been measured as 230 m (750 ft).

The queen, workers, winged reproductives, and larvae in the parent colony overwinter in a metabolic state termed diapause. In temperate regions, diapause is a period of dormancy during which the ants are in a state of "suspended animation." The wood encasing the colony insulates it from cold temperatures. In addition, larvae, workers, and reproductives have glycerol in their hemolymph (blood) that acts as antifreeze (Cannon and Fell 1992).

Colonies in temperate regions break diapause from January to June (depending on the latitude, elevation, and habitat), and the queen begins her first egg-laying of the season; *C. modoc* queens may lay eggs for 7 to 10 days (Hansen and Akre 1985, 1990). The voracious appetite of the developing larvae triggers increased foraging activity on the part of the workers. The most intense foraging of the season coincides with increased food requirements of the rapidly developing larvae. A second shorter and less intense peak of activity occurs in June when the queen lays eggs for another 7 to 10 days. The

**Figure 2.10.** Trail constructed by carpenter ants (*Camponotus modoc*) through a lawn

colony enters diapause in September or October along with the late summer brood, which overwinter as larvae and complete development in late winter. Individual colonies may survive for more than 20 years (often with the same queen).

Since carpenter ants are primarily nocturnal, they rely on physical cues and chemical trails for orientation to and from the nest. Well-maintained physical trails and trunk lines serve as roadways through vegetation and debris (Fig. 2.10). Carpenter ants in extreme northern latitudes often travel underground, following natural hollows such as those left by decaying tree roots in the soil. Such tunnels are usually 1.5 to 3 cm (1–1.3 in) in diameter and may be as deep as 1 m (3.3 ft) below the surface (Hansen and Akre 1985). The distance traveled to obtain food varies. The surface trails of a *C. modoc* colony in California were 200 m (656 ft) long (David and Wood 1980), for example, and the tunnels used by a *C. herculeanus* colony in Ontario totaled 185 m (607 ft) in length (Sanders 1970).

Black carpenter ants (*C. pennsylvanicus*) have a distinct cycle of protein consumption that coincides with their brood production in the summer and fall (Cannon and Fell 2002). The quantity and quality of nitrogen in the protein and amino acids ingested are the key factors for growth and development (Hagen et al. 1984). Their consumption of carbohydrates, on the other hand, is relatively constant and more than twice as high as protein consumption (Cannon

**Figure 2.11.** Carpenter ants (*Camponotus modoc*) nesting under insulation

and Fell 2002). Carbohydrates are adult ants' primary energy source (Abbott 1978).

In structural infestations of carpenter ants, the parent colony is generally located outside in a tree, stump, stack of firewood, or landscape timbers. Nests in live trees are frequently located in hollows, heartwood, or dead limbs. Satellite colonies may occupy similar sites in one or more neighboring trees or may be in adjacent structures such as attic rafters, roof overhangs, bay windows, fascia boards, floor joists, box headers, wall voids, hollow curtain and shower rods, hollow doors or columns, spaces behind dishwashers and under cabinets, areas under or behind insulation in attics and crawl spaces, bath traps, and ceiling voids next to skylights and chimneys (Fig. 2.11). Parent colonies found inside structures are typically associated with a water leak or other source of moisture.

Houses built in woodland habitats are prime targets for carpenter ant infestations. Trees, landscape timbers, wooden porches and fences, and bay or box windows are all potential "hot spots." Homes with flat roofs, dormers, or hollow porch columns are potential sites for infestation as well. Leaky pipes or roofs, clogged gutters, and chimneys with improperly fitted flashing can create moisture problems, which are an open invitation for a carpenter ant infestation. Homes with multiple roof lines often suffer moisture damage in the attic

**Figure 2.12.** Excavated wood from a carpenter ant (*Camponotus modoc*) nest

if it is not properly sealed and adequately ventilated. Other conditions conducive to infestation include holes and cracks where utility lines enter the house, earth-to-wood contact (Fig. 2.12), and tree branches in contact with the building.

Carpenter ant control is expensive and often ineffective. Indeed, carpenter ants are the most common and problematic of all the ant pests in the United States (Granovsky 1990). In Europe, infestations are limited mainly to wooden structures in or near forests such as log cabins, barns, and hunting lodges, and poles and trees that are often associated with woodpecker damage, particularly the black woodpecker (*Dryocopus martius*) (Seifert 1996).

CONTROL

The inspection is the most important component of a successful control program for carpenter ants. Finding the nest(s) and foraging trails is essential, but carpenter ant nests are notoriously difficult to locate. Control is easiest to achieve when the parent colony and all the satellite colonies can be located (Hansen 2001). Night inspections are helpful in this regard, because many species of carpenter ants are nocturnal. Large numbers of ants emerge after sundown and disappear into the nest at sunrise (Nuss et al. 2005). Feeding the foraging carpenter ants diluted honey, sugary milk, and chopped insects such as crickets or mealworms and then following them on their homeward journey can help locate nests (Akre et al. 1994b).

Once the nests and trails have been located, residual treatments or baits are applied. Three insecticide formulations are commonly used in carpenter ant control (Hansen 2002): sprays, dusts, and baits.

*Sprays.* Nonrepellent insecticides such as fipronil and synthetic pyrethroids such as cyfluthrin and deltamethrin are effective as perimeter treatments when applied to the exterior foundation, around door and window frames, and under the lower edge of siding (Hansen 1989). Trails should also be treated. Locations prone to infestation (e.g., wooded lots) should receive a biannual perimeter treatment or at least one in the spring, when carpenter ants become active. These same insecticides can be applied around tree trunks harboring carpenter ants and to nests inside trees. Nonrepellent insecticides are effective at low concentrations and are readily transferred to other ants, thereby reducing the total population. Perimeter treatments are especially effective when satellite nests are indoors and the parent nest is outdoors. The interchange of ants between the nests spreads the active ingredient to the entire colony. Such treatments, however, usually do not result in total control of an infestation. Location and treatment of the parent colony should be the goal, with perimeter treatment assisting in reducing ant activity inside the building.

*Dusts.* A dust formulation of insecticide such as disodium octaborate tetrahydrate or deltamethrin can be applied in very light, thin layers to nests in structural voids or inside wood timbers. Dusts become repellent when applied too heavily. Compressed air or cordless electric dusters are helpful in treating wall voids because of their very light output and excellent coverage (Akre et al. 1995). Dusts can be injected into wall voids either through existing openings around plumbing and electrical lines or by drilling access holes.

*Baits.* Toxic baits typically take longer to achieve results than residual treatments because they rely on trophallaxis to distribute the toxicant through the colony (Fig. 2.13). In addition to potential points of entry into a structure, they are also useful for treating nests high in a tree or on neighboring property. Carpenter ants forage for insects and honeydew on the ground and in trees and other vegetation, and baits applied along foraging trails can be effective, particularly during the foraging season. A variety of baits should be offered to the ants to determine their preference, and bait stations should be monitored to determine acceptance, amount consumed, and reduction in the ant population

**Figure 2.13.** Carpenter ant (*Camponotus modoc*) workers feeding one another

(Hansen 2000). For example, collection of the protein Maxforce bait by *C. pennsylvanicus* foragers was significantly higher in spring than in fall (Tripp et al. 2000).

In an initial treatment for carpenter ants, a PMP may drill holes in wall voids and remove switch plates to apply dust in addition to applying a perimeter treatment outside (Akre et al. 1995). When baits are used in combination with residual treatments, the baits should be applied first, giving ants several days to feed on them and distribute the active ingredient throughout the colony.

Nonchemical techniques such as trimming back vegetation that provides access for ants onto buildings and sealing or caulking potential entry points into a structure are also helpful in preventing infestation.

### Key to Species of *Camponotus*

1   Mesosoma short, that of major worker no longer than head (mandibles excluded) (Fig. 2.14a); pronotum angular with margin edged by flat border; southern Florida, southern Texas, and Mexico . . . . . . . . . . . . . . Subgenus *Myrmobrachys* . . . . . . . . . . . . . . . . . . . . . . . . . *C. planatus*
    Mesosoma longer, that of major worker longer than head (mandibles excluded) (Fig. 2.14b); pronotum with rounded margin . . . . . . . . . . . . 2

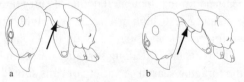

**Figure 2.14.** Profile of head and mesosoma. (a) *C. planatus* (arrow: angular pronotum) (b) *C. decipiens* (arrow: rounded pronotum)

2(1) Scapes and legs with numerous long, coarse, brownish or golden, erect hairs on all surfaces (Fig. 2.15a) . . . . . . . . . . . . Subgenus *Myrmothrix*
. . . . . . . . . . . . . . . . . . . . . . . . . . . . . . . . . . . . . . . . . . . . . . . *C. floridanus*

Scapes and legs with erect hairs that are fine, short, and usually whitish; hairs often confined to row of bristles on flexor surface of legs (Fig. 2.15b, c) . . . . . . . . . . . . . . . . . . . . . . . . . . . . . . . . . . . . . . . . . . 3

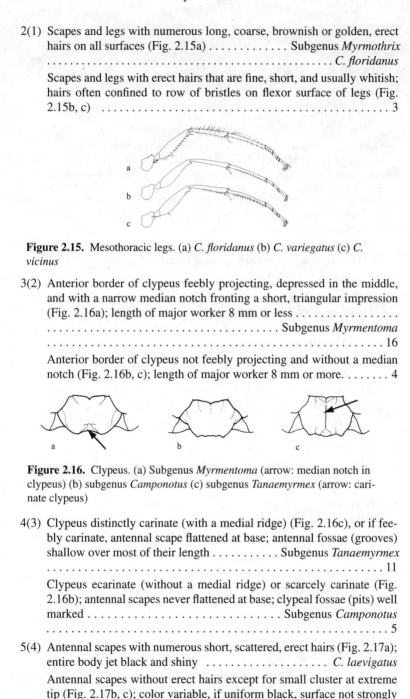

**Figure 2.15.** Mesothoracic legs. (a) *C. floridanus* (b) *C. variegatus* (c) *C. vicinus*

3(2) Anterior border of clypeus feebly projecting, depressed in the middle, and with a narrow median notch fronting a short, triangular impression (Fig. 2.16a); length of major worker 8 mm or less . . . . . . . . . . . . . . .
. . . . . . . . . . . . . . . . . . . . . . . . . . . . . . . Subgenus *Myrmentoma*
. . . . . . . . . . . . . . . . . . . . . . . . . . . . . . . . . . . . . . . . . . . . . 16

Anterior border of clypeus not feebly projecting and without a median notch (Fig. 2.16b, c); length of major worker 8 mm or more. . . . . . . . 4

**Figure 2.16.** Clypeus. (a) Subgenus *Myrmentoma* (arrow: median notch in clypeus) (b) subgenus *Camponotus* (c) subgenus *Tanaemyrmex* (arrow: carinate clypeus)

4(3) Clypeus distinctly carinate (with a medial ridge) (Fig. 2.16c), or if feebly carinate, antennal scape flattened at base; antennal fossae (grooves) shallow over most of their length . . . . . . . . . . Subgenus *Tanaemyrmex*
. . . . . . . . . . . . . . . . . . . . . . . . . . . . . . . . . . . . . . . . . . . . . 11

Clypeus ecarinate (without a medial ridge) or scarcely carinate (Fig. 2.16b); antennal scapes never flattened at base; clypeal fossae (pits) well marked . . . . . . . . . . . . . . . . . . . . . . . . . . . . Subgenus *Camponotus*
. . . . . . . . . . . . . . . . . . . . . . . . . . . . . . . . . . . . . . . . . . . . . 5

5(4) Antennal scapes with numerous short, scattered, erect hairs (Fig. 2.17a); entire body jet black and shiny . . . . . . . . . . . . . . . . . . . . *C. laevigatus*

Antennal scapes without erect hairs except for small cluster at extreme tip (Fig. 2.17b, c); color variable, if uniform black, surface not strongly shiny. . . . . . . . . . . . . . . . . . . . . . . . . . . . . . . . . . . . . . . . . . . . 6

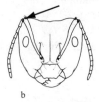
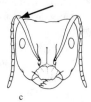

a                           b                           c

**Figure 2.17.** Anterior view of head with antennae. (a) *C. laevigatus* (arrow: numerous hairs) (b) *C. herculeanus* (arrow: hairs at tip only) (c) *C. modoc* (arrow: scape length surpasses posterior corner by more than diameter of scape)

6(5)  Antennal scapes of major worker reaching or barely surpassing posterior corners of head in face view (Fig. 2.17b); epinotum dark red, rest of ant dull black . . . . . . . . . . . . . . . . . . . . . . . . . . . . . . . . . . . *C. herculeanus*

Antennal scapes of major worker surpassing posterior corners of head in face view by an amount greater than their maximum diameter (Fig. 2.17c); color not as above . . . . . . . . . . . . . . . . . . . . . . . . . . . . . . . 7

7(6)  Pubescence on gaster absent or very fine and sparse (Fig. 2.18a), entire surface of gaster distinctly shiny. . . . . . . . . . . . . . . . . . . . . . . . . . . . . 8

Pubescence on gaster coarse and dense (Fig. 2.18b, c), surface of gaster dull except for narrow band at posterior edge of each segment . . . . . 9

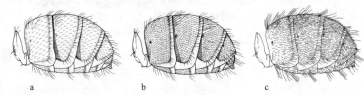

a                           b                           c

**Figure 2.18.** Gaster. (a) *C. novaeboracensis* (b) *C. modoc* (c) *C. pennsylvanicus*

8(7)  Punctures on head coarse and conspicuous; mesosoma red, head and gaster brownish black . . . . . . . . . . . . . . . . . . . . . . . *C. novaeboracensis*

Punctures on head fine and inconspicuous; color variable but mesosoma never red . . . . . . . . . . . . . . . . . . . . . . . . . . . . . . . . . . *C. americanus*

9(7)  Pubescence on gaster less than half as long as erect hairs (Fig. 2.18b); western species. . . . . . . . . . . . . . . . . . . . . . . . . . . . . . . . . . . . *C. modoc*

Pubescence on gaster about as long dorsally as erect hairs (Fig. 2.18c); eastern species . . . . . . . . . . . . . . . . . . . . . . . . . . . . . . . . . . . . . . . 10

10(9)  Head, mesosoma, pedicel, and gaster dull black; pubescence pale yellow or white . . . . . . . . . . . . . . . . . . . . . . . . . . . . . . . . . . . *C. pennsylvanicus*

Posterior portion of mesosoma, pedicel, and base of first gastral segment bright reddish brown, pubescence golden yellow . . . . *C. chromaiodes*

11(4)  Middle and hind tibiae without row of graduated, erect bristles on their
       flexor surfaces (Fig. 2.15b). . . . . . . . . . . . . . . . . . . . . . . . . . . . . . . . 12

       Middle and hind tibiae with row of graduated, erect bristles on their
       flexor surfaces (Fig. 2.15c). . . . . . . . . . . . . . . . . . . . . . . . . . . . . . . 13

12(11) Entire body yellow, with blackish bands on gastral segments (Fig. 2.19);
       Hawaii or isolated areas on West Coast . . . . . . . . . . . . . *C. variegatus*

       Body rusty red-brown, head darker than mesosoma; gaster dark brown
       or black (southern Florida). . . . . . . . . . . . . . . . . . . . . . *C. tortuganus*

**Figure 2.19.** Gaster of *C. variegatus*

13(11) Scape of major worker distinctly flattened at base and with flattened por-
       tion forming a small lateral lobule (small lobe) (Fig. 2.20a) . . . . . . . . .
       . . . . . . . . . . . . . . . . . . . . . . . . . . . . . . . . . . . . . . . . *C. semitestaceus*

       Scape of major worker not flattened at base, or if flattened, without a lat-
       eral lobule (Fig. 2.20b) . . . . . . . . . . . . . . . . . . . . . . . . . . . . . . . 14

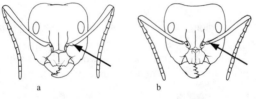

**Figure 2.20.** Anterior view of the head with antennae. (a) *C. semitestaceus*
(arrow: lateral lobule) (b) *C. vicinus* (arrow: no lateral lobule)

14(13) Genae strongly shining with very small, inconspicuous punctures; color
       uniform chestnut brown . . . . . . . . . . . . . . . . . . . . . . . . . *C. castaneus*

       Genae feebly shining or dull, the punctures coarser and conspicuous;
       color not uniformly brown . . . . . . . . . . . . . . . . . . . . . . . . . . . . . 15

15(14) Scape of major worker flattened at base; genae without erect hairs
       . . . . . . . . . . . . . . . . . . . . . . . . . . . . . . . . . . . . . . . . . . . . . . *C. vicinus*

       Scape of major worker not flattened at base, genae with erect hairs
       . . . . . . . . . . . . . . . . . . . . . . . . . . . . . . . . . . . . . . . . . . . *C. acutirostris*

16(3)  Mesosomal profile distinctly depressed at metanotal suture (Fig. 2.21a)
       . . . . . . . . . . . . . . . . . . . . . . . . . . . . . . . . . . . . . . . . . . . . . . *C. hyatti*

       Metanotal suture not depressed (Fig. 2.21b, c). . . . . . . . . . . . . . . . . 17

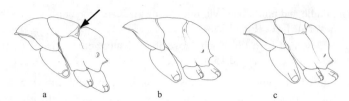

**Figure 2.21.** Mesosomal profile. (a) *C. hyatti* (arrow: depressed metanotal suture) (b) *C. clarithorax* (c) *C. discolor*

17(16)  Malar area with conspicuous suberect to erect, short hairs arising from coarse, elongate foveae (depressions or pits) (Fig. 2.22a, b). . . . . . . 18

Malar area without suberect or erect hairs, except sometimes near base of mandible (Fig. 2.22c). . . . . . . . . . . . . . . . . . . . . . . . . . . . . . . . . 21

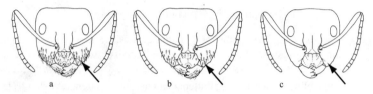

**Figure 2.22.** Anterior view of the head. (a) *C. caryae* (arrow: with erect hairs) (b) *C. subbarbatus* (arrow: with erect hairs) (c) *C. essigi* (arrow: without erect hairs)

18(17)  Clypeus with long, erect hairs along margins and numerous shorter hairs across disc (Fig. 2.22a) . . . . . . . . . . . . . . . . . . . . . . . . . . . . . . . . . . 19

Clypeus with long, erect hairs along margins and adjacent to clypeus, and fewer than three hairs across disc (Fig. 2.22b) . . . .*C. subbarbatus*

19(18)  Epinotal profile rounded (Fig. 2.21c) . . . . . . . . . . . . . . . . . . . . . . . . 20

Epinotal profile angular (Fig. 2.21b) . . . . . . . . . . . . . . . *C. clarithorax*

20(19)  Erect hairs of clypeus of varying lengths, with shortest about as long as those of malar area; more erect hairs on malar area (Fig. 2.23a) . . . . . .
. . . . . . . . . . . . . . . . . . . . . . . . . . . . . . . . . . . . . . . . . . . . . . . . . .*C. caryae*

Erect hairs of clypeus distinctly long or short, short hairs shorter than those of malar area; fewer erect hairs on malar area (Fig. 2.23b)
. . . . . . . . . . . . . . . . . . . . . . . . . . . . . . . . . . . . . . . . . . . . . . . . .*C. discolor*

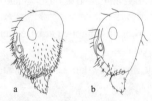

**Figure 2.23.** Profile of the head. (a) *C. caryae* (b) *C. discolor*

21(17)  Epinotum in profile with basal face flat or nearly so, almost entirely on same plane as the mesonotum, abruptly rounded or subangulate at juncture with posterior slope of epinotum (Fig. 2.24a) . . . . . . . . . . . . . 22

Epinotum in profile curved or straight, but sloping toward broadly rounded juncture with posterior slope of epinotum (Fig. 2.24b)  . . . 23

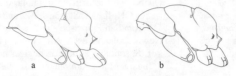

a                                    b

**Figure 2.24.**  Profile of the mesosoma. (a) *C. essigi* (arrow: basal face of epinotum flat) (b) *C. decipiens* (arrow: basal face of epinotum rounded)

22(21)  Head, mesosoma, and appendages rusty red-brown, gaster black; lower malar area and frontal lobes dull, densely tessellated (checkered) . . . . . . . . . . . . . . . . . . . . . . . . . . . . . . . . . . . . . . . . . . . . . . . . . . . . *C. sayi*

Color variable but often wholly blackish; lower malar area and frontal lobes subpolished to shiny. . . . . . . . . . . . . . . . . . . . . . . . . . . . *C. essigi*

23(21)  Head and mesosoma dark red to black, gaster entirely black . . . . . . . . . . . . . . . . . . . . . . . . . . . . . . . . . . . . . . . . . . . . . . . . . . . . . . . . *C. nearcticus*

Head, mesosoma, and appendages red to yellowish red, gaster brown . . . . . . . . . . . . . . . . . . . . . . . . . . . . . . . . . . . . . . . . . . . . . . *C. decipiens*

## Thatching, Wood, and Field Ants

*Formica* species (Plate 1c)

### IDENTIFYING CHARACTERISTICS

Workers of this genus range in length from 3 to 9 mm and may vary in color from all black or brown to bicolored red and black. The single segmented petiole has a distinct vertical node, and the acidopore is circular, terminal, and fringed with hairs. The workers of a given species are polymorphic, with the majors approaching the size of small carpenter ants. They can be distinguished from carpenter ants by the dorsal profile of the mesosoma, which in *Formica* is uneven or notched.

### DISTRIBUTION

*Formica* is a large, complex, holarctic genus with more than 150 species worldwide (Snelling and George 1979; Czechowski 2002). More than 78

**Table 2.2.** Generalized distribution pattern of potential pest species of *Formica* in North America and Europe

| Group | Species | Approximate geographic distribution |
|---|---|---|
| *Exsecta* | *exsectoides* | Nova Scotia south to GA, west to Ontario, WI, and IA; south to CO and northern NM |
| *Fusca* | *argentea* | Quebec west to British Columbia, south to SC, OH, IL, IA, SD, NM, AZ, and CA |
| | *francoeuri* | CA from San Francisco south to Mexico |
| | *fusca* | Newfoundland west to AK, northern half of U.S. south to latitude 38 degrees north except in larger mountain ranges, where it extends farther south; Europe |
| | *neorufibarbis* | Newfoundland west to AK and south to MA, MI, MN, SD, NM, AZ, and CA |
| | *subsericea* | New Brunswick and Quebec south to FL and west to Manitoba, MT, IA, KS, MO, MS |
| *Microgyna* | *densiventris* | CO, NM, and UT west to CA and north to WA |
| *Neogagates* | *perpilosa* | WY, CO, KS, OK, and TX west to CA and Mexico |
| *Pallidefulva* | *nitidiventris* | Southern Quebec and Ontario south to the mountains of northern GA and west to WI and IA; sporadic in foothills regions in WY, SD, CO, NM, and TX |
| *Rufa* | *integroides* | Coastal mountains of CA and western slopes of the Sierra Nevada north to WA |
| | *obscuripes* | Northern IL and WI northwest to WA and British Columbia, extending south down the Rocky Mts. to AZ and NM |
| | *obscuriventris* | Southern Canada and New England south to GA and west to WI and the Black Hills of SD, sporadic at high elevations in southern Rocky Mts. |
| | *oreas* | Saskatchewan and ND to NM, northwestward to WA and southern Alberta |
| | *planipilis* | ND, CO, NV, WY, UT, ID, OR, WA, and British Columbia |
| | *polyctena* | Europe |
| | *pratensis* | Europe |
| | *propinqua* | Central CA north to WA along eastern slopes of Sierra Nevada and Cascade Mts., CO and UT |
| | *ravida* | ND, SD, CO west to British Columbia, WA, and CA |
| | *rufa* | Europe |
| | *subnitens* | ND, WY, CO, NM, British Columbia, OR, and CA |
| | *truncorum* | Europe |
| *Sanguinea* | *aserva* | Newfoundland west to Yukon Territory and AK, south to NY, MN, ND, CO, NM, AZ, and CA |
| | *subintegra* | Newfoundland, Nova Scotia, and Ontario south to SC and TN, and west to ND, IA, and KS |

*Sources:* Creighton 1950; Wheeler and Wheeler 1986; Seifert 1996; Czechowski et al. 2002; Hedlund 2003.

species occur in North America in a wide variety of habitats ranging from open fields to montane environments and at altitudes ranging from sea level to above 4267 m (14,000 ft). The *Formica* species may become nuisance pests when they occur in proximity to human habitation (see Table 2.2 for geographic distribution).

The forests of Europe are home to several species of *Formica* including *F. rufa, F. polyctena, F. truncorum,* and *F. pratensis.* The latter species is more common in southern Europe and prefers to nest in dry areas of open forests, meadows, and steppes.

## BIOLOGY AND HABITS

Members of this genus are called field ants because some species prefer to nest in open areas (Cook 1953), but many northern species also occupy woodland habitats (Snelling and George 1979). Some species are known as thatching or mound-building ants because they construct their nests from small twigs, grass stems, leaves, and pine or fir needles (Akre 1992). The mounds they build may be spectacular; for example, an *F. polyctena* colony may be 3 m (9.8 ft) in diameter and 1.5 m (4.5 ft) tall. Colonies may contain more than a million workers and a thousand queens. Large colonies of *F. polyctena* may consist of several satellite nests and encompass large areas, and their foragers can be found up to 100 m (328 ft) away from the nest.

The mounds are built to capture light rays (Fig. 2.25). The temperatures within them vary according to depth and time of day, and the ants continually move brood from place to place to keep them at an optimal temperature. The mounds may blemish lawns or golf courses, and in the case of the Allegheny mound ant (*F. exsectoides*) of the northeastern United States are similar in size to those of the red imported fire ant and have caused "fire ant scares" in that part of the country (Bennett et al. 1997). *Formica* species are common around homes and other buildings but are rarely found indoors. When mounds are located near a foundation, however, ants may enter under the lower edge of the siding and nest in a wall void (Fig. 2.26). They also may enter structures to feed on sweet materials, but their natural diet consists of honeydew, nectar, and arthropods.

Some species are injurious to seedling trees and other plants and have been known to damage buds of fruit and other deciduous trees in the spring. Many species tend aphids, herding them on trees, shrubbery, and other plants. The ants protect the aphids from their natural predators, and an aphid infestation may be the result. Most species are territorial and patrol foraging trails on tree trunks and surrounding areas. Their mandibles are sharp enough to pierce the

**Figure 2.25.** *Formica* spp. nest with mound of thatching material

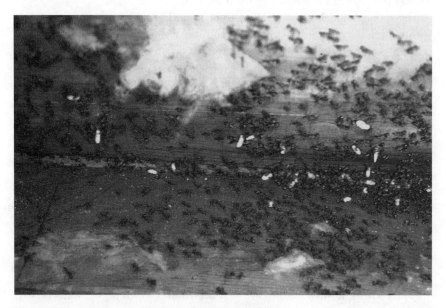

**Figure 2.26.** *Formica* spp. nesting under insulation in a house

skin, and they may spray formic acid into the wound, causing a painful but generally harmless reaction. The formic acid produced by ants on a thatching mound is sometimes produced in large enough quantities to be detected when approaching a nest.

Most species of *Formica* are polygynous and some are polydomous (a single colony with many nest sites), although even colonies belonging to a single species may vary in these characters. For example, *F. rufa* colonies are monogynous in Europe except in the British Isles, where nests may contain up to one hundred queens (Czechowski et al. 2002).

Colonies are established in a variety of ways depending on the species. Some species practice temporary parasitism in which a queen of one species usurps a nest belonging to another species; for example, young queens of *F. rufa* and *F. polyctena* parasitize the nests of *F. fusca* and *F. rufa,* respectively. Polygyne colonies of *F. rufa* may also multiply by fission or budding. Slave-making species rob brood from the nests of other species and rear the workers as slaves. For example, *F. subsericea* workers may become slaves of *F. subintegra,* and *F. argentea* and *F. fusca* workers are sometimes enslaved by other species of *Formica.*

## CONTROL

*Formica* ants should generally be considered beneficial because of the number of defoliating insects they consume. If they become a nuisance to a homeowner because they are invading a structure or because they are biting, the entire surface of the mound or nest can be treated with a residual spray, dust, or granule formulation. The treatment can be enhanced by digging into the nest with a sturdy shovel while applying a recommended amount of insecticide. It is important to follow the directions on the label, as some treatments require application of water to carry the insecticide into the nest. Nests should be checked for activity 7 to 10 days later.

Baits have been used to control pest species of *Formica* in agricultural and wildlife areas. An experimental anchovy-based granular bait with 0.005% imidacloprid as the active ingredient significantly reduced foraging activity of *F. perpilosa* in grape vineyards, and Advance ant bait with abamectin was used to control *F. francoeuri* at a least tern breeding site. Baits are thus potentially useful for field ant control, but extensive research is still needed to determine bait preference and efficacy.

*Formica* species are protected in some European countries because of their ecological importance, and since they generally do not invade homes, control measures are not recommended.

## Key to Groups of *Formica*

1   Ventral border of clypeus with middle notch; usually bicolored; epinotum
    short and angulate in profile (Fig. 2.27a) . . . . . . . . . . . *sanguinea* group
    Ventral; border of clypeus without notch; color variable (Fig. 2.27b) . . 2

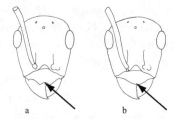

**Figure 2.27.** Ventral border of clypeus. (a) *F. aserva* (arrow: median notch on
clypeus) (b) *F. perpilosa* (arrow: no median notch on clypeus)

2(1)  Slender; epinotum rounded in profile (Fig. 2.28a). . . . . . . . . . . . . . . . . 3
      Robust; epinotum angulate in profile (Fig. 2.28b). . . . . . . . . . . . . . . . 4

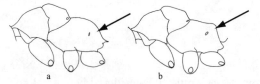

**Figure 2.28.** Epinotum. (a) *F. nitidiventris* (arrow: rounded epinotum) (b) *F.
exsectoides* (arrow: angular epinotum)

3(2)  Scape slender, long (more than 1.25 times length of head); barely curved
      at base; frontal carina sigmoid (Fig. 2.29a) . . . . . . . . *pallidefulva* group
      Scape shorter (less than 1.25 times length of head); frontal carina subpar-
      allel (Fig. 2.27b) . . . . . . . . . . . . . . . . . . . . . . . . . . . . . . *neogagates* group

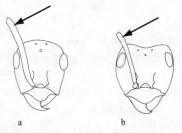

**Figure 2.29.** Face view. (a) *F. nitidiventris* (arrow: longer scape) (b) *F. exsectoides* (arrow: shorter scape)

4(2)  Occipital border distinctly concave; pronotum in profile with basal and de-
      clivous faces meeting at an angle (Fig. 2.29b, Fig. 2.30a) . . . . *exsecta* group

Occipital border at most slightly concave, usually flat; pronotum in profile
evenly convex, not angulate (Fig. 2.30b) ........................ 5

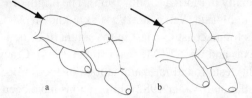

**Figure 2.30.** Pronotum and mesonotum profile. (a) *F. exsectoides* (arrow: angular pronotum) (b) *F. fusca* (arrow: rounded pronotum)

5(4) Bicolored, head and thorax reddish and lighter than gaster, or if infuscated
(smokey gray-brown), the infuscation not completely masking reddish
ground color; gaster brown or black ............................ 6

Black, brown, or bicolored; if bicolored, thorax lighter than gaster and upper portion of head. .............................. *fusca* group

6(5) Erect hairs on pronotum of worker broadly spatulate (rounded at the tips)
............................................... *microgyna* group

Erect hairs on pronotum of worker not spatulate ........... *rufa* group


# Crazy Ant and Related Species

*Paratrechina* species (Plate 1d)

### IDENTIFYING CHARACTERISTICS

Workers of the crazy ant, *P. longicornis,* are slender, fast-moving ants with long
legs and antennae (Smith 1965). They range in length from 2.3 to 2.9 mm
(Trager 1984) and have a dark brown to black body (Smith 1965) with a bluish
iridescence (Snelling and George 1979). Workers of *P. bourbonica* resemble
crazy ants but are more robust and have shorter legs and antennae (Hedges
1997). They are dark brown to nearly black, 2.6 to 3.2 mm long, and their antennal scape has many conspicuous hairs (Trager 1984). Workers of *P. vividula* are
2 to 2.5 mm long and have a strongly shiny appearance with a brown head, dark
yellowish brown thorax, and dark brown gaster (Wheeler and Wheeler 1986);
*P. pubens* workers are 2.55 to 3.1 mm long and reddish brown (Trager 1984).


### DISTRIBUTION

Crazy ants are invasive and tramp species that probably originated in Asia or
Africa (McGlynn 1999; Morgan et al. 2005) and have been widely distributed

by commerce (Snelling and George 1979). They are found sporadically throughout the United States (Bennett et al. 1997) and are common along the Gulf Coast, especially in Florida (Smith 1965). The origin of *P. bourbonica,* a major urban pest in Florida (Vail et al. 1994), is unknown (McGlynn 1999); *P. vividula,* which is found coast to coast in the southern United States, is probably native to Texas and western Mexico (Trager 1984). Crazy ants were the most common ant pests found in examinations of food-processing plants in south Texas (Shetlar and Walter 1982). Infestations have been found in a hotel kitchen in San Francisco (Shetlar and Walter 1982) and sporadically from Los Angeles to San Diego (Snelling and George 1979). Biosphere 2 outside Tucson, Arizona, became infested with them in the early 1990s (Wetterer et al. 1999).

*Paratrechina vividula* is common in many urban locations in California (Ward 2005). Wheeler and Wheeler (1986) collected *P. vividula* at sixteen sites in Nevada, and all were associated with man-made structures. *Paratrechina pubens,* which originated in the Neotropics, is established outdoors in Miami, Florida, and is occasionally found farther north (e.g., Washington, D.C.) in greenhouses (Trager 1984). Structural infestations have been reported in a hospital in Miami, a house in Boca Raton, and around a commercial building near Homestead, Florida (Klotz et al. 1995).

In colder northern regions, crazy ants can be year-round pests indoors (Smith 1965). *Paratrechina vividula* and *P. longicornis* were intercepted in Wismar, Germany, in a shipment of fruit from east Africa. They were still onboard a freighter when they were found and were eradicated before they reached land (Steinbrink 2000). Between 1995 and 2005, there were two reports of *P. longicornis* in Switzerland (Umwelt-und Gesundheitsschutz Zürich 2004). Crazy ants have also been transported to Australia, New Zealand, India, Southeast Asia, Central and South America, the Arabian Peninsula, and several Atlantic and Pacific islands, including Hawaii (McGlynn 1999).

## BIOLOGY AND HABITS

These ground-dwelling ants occasionally become pests in homes and greenhouses (Creighton 1950) and are sometimes referred to as "sugar ants" because of their dietary preference for sweets (Smith 1965). Most notorious among the introduced *Paratrechina* species is the crazy ant (*P. longicornis*) which derives its common name from its erratic, jerky movements (Thompson 1990). Individuals follow pheromone trails produced by the hindgut of foragers (Blum and Wilson 1964), sometimes over long distances (Morgan et al. 2005). Aggression between different field-collected colonies of crazy ants indicates that they are not unicolonial (Lim et al. 2003).

The colonies that Blake (1940) examined were small, with two thousand or so workers and eight to forty queens each, but Hedges (1997) found colonies numbering tens of thousands living in piles of debris and within landscape mulch in Florida and Texas. It is not unusual for a colony to leave its nest and move to another site (Blake 1940). Nuptial flights are abortive, with males waiting outside the nest to mate with emerging dealated queens (Trager 1984).

Crazy ants nest in both dry and moist environments in trash, plants, rotten wood, and soil (Smith 1965). They also nest in landscape mulch and behind thick vegetation adjacent to foundations; indoor nests are located in wall voids, under carpeting, and in potted plants (Hedges 1997). In Florida, they prefer nesting in cracks in masonry, under flowerpots, and in the crowns of palm trees (J. Warner, pers. comm., 2004).

*Paratrechina* species are omnivores that feed on honeydew, both live and dead insects, and various household foods, especially sweets (Smith 1965). The *P. longicornis* that became the predominant ant species in Biosphere 2 fed primarily on honeydew produced by mealybugs and scales (Wetterer et al. 1999). In food preference tests, tuna, Xstinguish (protein + carbohydrate bait), and sugar water were highly attractive (Stanley and Robinson 2007).

## CONTROL

The Cooperative Extension Service in Florida, where crazy ants are major household pests, recommends sugar-based baits in spring and fall, and protein-based baits in summer (Nickerson and Barbara 2000). Nonchemical measures such as good sanitation practices, caulking, and sealing potential entry points are also important for efficient control.

Perimeter sprays are most effective when combined with treatments of nests on the property (Hedges 1998). One PMP in Florida who treats crazy ants on a regular basis follows trails to the nesting areas, sprays these, and then returns to spray the trails and foraging areas. Nests located in palm crowns or other protected areas that cannot be sprayed are treated with sweetened liquid baits (J. Warner, pers. comm., 2004).

### Key to Species of *Paratrechina*

1 Scapes and legs unusually long; weakly shiny black or gray with sparse, short pubescence . . . . . . . . . . . . . . . . . . . . . . . . . . . . . . . . . *P. longicornis*
   Scapes and legs of usual proportions; variously colored, shiny or covered with dense pubescence . . . . . . . . . . . . . . . . . . . . . . . . . . . . . . . . . . . 2

2(1)  Mesosoma and gaster covered with pubescence, or pubescence lacking
      only on pleura and pronotum; surface dull .................... 3
      Thorax and gaster with greatly reduced pubescence, shiny .... *P. vividula*

3(2)  Body light reddish brown with light brown pilosity (hairs) .... *P. pubens*
      Body dull brown to black, pilosity short, stout, and dark ... *P. bourbonica*

## "Small or False Honey Ant" or "Winter Ant"

*Prenolepis imparis* (Plate 1e)

### IDENTIFYING CHARACTERISTICS

Workers are monomorphic, 2 to 4 mm long, and have a one-segmented petiole. The antenna has twelve segments and lacks a club. A constriction of the
mesothorax gives the thorax an hourglass shape when viewed from above (Vail
et al. 1994), and the gaster is triangular and wider than the head (Wegner 1991).
The body is shiny and varies in color from light brown to dark brown or black.

### DISTRIBUTION

The small honey ant is found throughout the United States, sporadically in
southern Canada, and in central Mexico, often associated with oak trees
(Gregg 1963; Ebeling 1975; Wheeler and Wheeler 1986).

### BIOLOGY AND HABITS

As one of its common names suggests, the winter ant is cryophilic and begins
foraging at temperatures around freezing (Talbot 1943; Creighton 1950). During summer, these ants estivate for several weeks and forage very little or not
at all (Talbot 1943). Mating flights take place early in the spring, usually from
March to April (Smith 1965), when other ants are just emerging from hiber-
nation (Talbot 1943). In Florida, their seasonal activity is restricted to November to April, and most colonies are polygynous and contain 600 to 10,300
workers (Tschinkel 1987). Farther north, monogynous colonies that rarely exceed a few thousand individuals are more common (Smith 1965), although
larger colonies have been observed in Oregon foraging on flowering camellias
in early spring.

  *Prenolepis imparis* nests are usually in damp soil in shady places, seldom
under stones or other objects (Smith 1965), and a crater of small soil pellets
surrounds the central entrance (Thompson 1990). Nests measured in Florida
were 2.5 to 3.6 m (8–12 ft) deep, several times the depth of nests farther north,

and consisted of a series of chambers connected to a single vertical tunnel (Tschinkel 1987).

Workers commonly invade houses to forage, and occasionally to nest, as evidenced by alates sometimes found indoors (Smith 1965). Plants in contact with structures offer the ants an opportunity to enter. They will also nest beneath slab foundations and enter buildings through expansion joints or cracks (Wegner 1991).

The workers forage on live and dead insects, honeydew, plant sap, and fruit (Creighton 1950; Smith 1965). In houses, they feed on all types of sweets, but also on bread and meat (Smith 1965). Corpulent workers engorged with fat provide nutrients for the brood that the colony produces once a year (Tschinkel 1987).

CONTROL

Application of residual insecticides to nests is the recommended treatment, although sweet liquid baits containing boric acid are also effective (Wegner 1991; Hedges 1998). Dust formulations of boric acid or diatomaceous earth are effective for nests in voids (Wegner 1991).

## Cornfield, Moisture, Black Garden, and Citronella Ants

*Lasius* species (Plate 1f)

IDENTIFYING CHARACTERISTICS

Workers are small (approximately 2–5 mm depending on the species) and monomorphic. They have a well-defined mesonotal suture and round epinotal spiracles. The frontal carinae are feebly developed, and the maxillary palps are long with six segments. Ocelli are small or absent. The body color varies from yellow to black (Wilson 1955; Wheeler and Wheeler 1986): *L. alienus, L. niger, L. platythorax, L. brunneus, L. pallitarsus,* and *L. neoniger* are light to blackish brown, rarely yellowish brown; *L. neglectus* is yellowish brown; *L. flavus* has a brown head and reddish yellow thorax and gaster (and also has reduced eyes and shortened maxillary palps); *L. subumbratus* and *L. umbratus* are reddish yellow; *L. fulginosus* is jet black; and *L. claviger, L. interjectus, L. murphyi,* and *L. latipes* vary from pale yellow to yellowish red (Smith 1965). Species of the subgenus *Acanthomyops* emit a distinct lemonlike odor (oil of citronella) from their mandibular glands when alarmed and have short, three-segmented maxillary palps; other members of this genus have maxillary palps with six segments. The Subgenus *Acanthomyops* has been embedded in *Lasius* (Ward 2005).

DISTRIBUTION

Species of *Lasius* are distributed discontinuously throughout North America and Europe (see Table 2.3 for geographic distribution). Those with the widest

**Table 2.3.** Generalized distribution patterns of pest species of *Lasius* in North America and Europe

| Subgenus | Species | Approximate geographic distribution |
|---|---|---|
| *Acanthomyops* | *claviger* | Southern New England west to MN and IA, and south to TN and NC |
| | *interjectus* | Coast to coast in the northern U.S., extending south down the Rocky Mts. to NM and in the Appalachian Mts. to the eastern Gulf states; abundant only in the central and northeastern states |
| | *latipes* | Coast to coast in the northern U.S. and extreme southern Canada; south down the Rocky Ms. to northern NM; rare south of PA in the eastern U.S. |
| | *murphyi* | Coast to coast in extreme southern Canada and the northern U.S.; extending south down the Rocky Mts. to NM and in the Appalachian Mts. to GA |
| *Cautolasius* | *flavus* | From Nova Scotia to NC and AL, and westward to the Pacific; Europe |
| *Chthonolasius* | *subumbratus* | Nova Scotia and ME west to Saskatchewan, WA, and OR, south to NM and AZ |
| | *umbratus* | Nova Scotia, New Brunswick, and Quebec south to FL and west to ID, UT, and AZ; southern and central Europe |
| *Dendrolasius* | *fuliginosus* | Southern and central Europe |
| *Lasius* | *alienus* | Southern Canada and the lower 48 U.S. states; Europe |
| | *brunneus* | Central and southern Europe |
| | *emarginatus* | Southern Europe |
| | *neoniger* | From southern ME across southern Canada to ID, south to CA and NM and a line joining the northern border of the OK panhandle to the FL panhandle |
| | *niger* | Southern AK, coast to coast in southern Canada and the northern U.S.; extending south in the Appalachian and Rocky mts. and the Sierra Nevada; Europe |
| | *pallitarsus* | From Nova Scotia and Quebec to MA, west through ND, SD, and southern Alberta, south through NM, west to the Pacific, and north through British Columbia to AK |
| | *paralienus* | Western and central Europe |
| | *platythorax* | Europe |
| | *psammophilus* | Central and northern Europe |
| | *neglectus* | Central and southern Europe |

*Sources:* Creighton 1950; Wheeler and Wheeler 1986; Seifert 1996; Czechowski et al. 2002; Hedlund 2003.

distribution are *L. alienus* and *L. flavus,* which are found in both the United States and Europe. The European species, *L. niger,* was split into sibling species: *L. niger* is found in open areas, and *L. platythorax* occupies woodland habitats (Seifert 1992). The invasive garden ant (*L. neglectus*), an introduced species in western Mediterranean areas and central Europe, probably originated in Asia Minor (Rey and Espadaler 2004). Species formerly known as *Acanthomyops* are found in North America (Smith 1965).

BIOLOGY AND HABITS

The name "cornfield ant" derives from the mutualistic relationship between *L. neoniger* and corn root aphids (Smith 1965). The ants store aphid eggs over winter and in spring carry the newly hatched aphids to the roots of grasses and then later to the roots of corn. This ant may also tend aphids on cotton and wheat. Cornfield ants nest in open areas in the soil or beneath stones. Other *Lasius* species nest in decayed logs and stumps (Fig. 2.31). The name "moisture ant" is applied to species that nest in decayed wood (Akre and Antonelli 1992). These ants commonly nest in rotting wood in houses and in form lumber buried in the soil next to foundations (Fig. 2.32). Moisture ants construct galleries from the rotting wood and may cement together fragmented pieces of

**Figure 2.31.** *Lasius fuliginosus* nest in the trunk of a tree

**Figure 2.32.** Moisture ant (*Lasius pallitarsus*) workers nesting in wood

**Figure 2.33.** Carton material of *Lasius* spp. from a nest under a board in a house

**Figure 2.34.** Paper nest of *Lasius fuliginosus* found below a wood floor in East Germany

decayed wood into carton-like material (Fig. 2.33). The carton in *L. fulginosus* nests is composed of macerated wood and soil particles hardened with honeydew and secretions from the ants' mandibular glands; the mixture is stabilized with mycelia of a fungus from the *Ascomycota* group (Fig. 2.34; Stumper 1950 cited by Wilson 1955; Seifert 2007). The galleries may be confused with galleries made by subterranean termites, but those of the latter are more linear and tubelike.

*Lasius neoniger* and the "European black garden ant" (*L. niger*) nest outdoors in areas with sandy soil and low vegetation. In moist soil with heavy ground vegetation, mounds of *L. niger* are designed to exceed the height of the grass (Wilson 1955) and are particularly visible after rainfall. Flat stones and terrace slabs create preferred nesting sites around buildings because they protect the colonies from rain and serve as heat reservoirs. Warm temperatures associated with composting wood and other decaying vegetation are also beneficial for a growing colony. Nests constructed below sidewalks often cause subsidence and accumulation of soil on the pavement.

Nests of "black garden ants" consist of a network of connected vertical and horizontal tunnels that may cover several square meters. The queen frequently moves to a new chamber. Although the nest sites of different colonies are often linked, each colony and its foraging trails are distinct.

"Black garden ants" tend aphids and other homopterans for honeydew, and also eat living and dead insects. Indoors, they are attracted to sugar-based products like marmalade, fruit juice, and honey. Workers are also attracted to meat, cheese, and other protein-containing substances when brood is developing in the nest. Foragers follow pheromone trails that may extend more than 30 m (98 ft).

Swarming occurs in late summer and early autumn. The large (8 mm) brown females and smaller (4 mm) dark males develop once a year and initiate mating flights, with males leaving the nest a few hours earlier than females. After mating, an inseminated female starts a new colony by building a nest chamber beneath a stone or bark, or in some other suitable area. She produces eggs that develop into larvae in the fall; the larvae overwinter in the nest and complete their development in the spring. A colony will grow to several thousand ants by the second year and may have up to 100,000 ants after 10 years.

"Citronella ants" nest in exposed soil, under objects such as stones, and in rotting wood. *Lasius interjectus* and *L. latipes* may create large mounds 0.3 m (ca. 1 ft) or more in diameter (Smith 1965). Alates of most species overwinter in nests and leave on mating flights in the spring, although reproductives of *L. claviger* may swarm in late summer and fall (Smith 1965). Fertilized females may establish nests independently or as temporary parasites of other *Lasius* species. The workers feed almost exclusively on honeydew collected from subterranean root aphids and mealybugs.

"Citronella ants" become a nuisance when the alates swarm or the workers dump soil from cracks in the floor or basement walls (Smith 1965). The overwintering alates emerge during the late winter or spring from cracks and crevices in walls and floors or from the basement and foundation, and many homeowners mistake them for termites.

*Lasius neoniger* is a significant pest of golf courses because colonies build mounds of soil on putting greens and tee areas. Nests consist of a series of shallow, interconnected chambers that lead to the main nests in natural soils of adjacent rough areas (Maier and Potter 2005).

In several urban areas of Europe the invasive garden ant (*L. neglectus*), which forms large supercolonies, attains pest status comparable to that of Argentine ants (Rey and Espadaler 2004). In a suburb near Barcelona, for example, one population occupied more than 14 ha. In addition to being a nuisance pest when they enter homes, the ants also chew on electrical wires and can short-circuit electrical appliances (Fig. 2.35).

**Figure 2.35.** Nest of black garden ant (*Lasius niger*) in the electrical part of an outdoor light

## CONTROL

The *Lasius* species are minor structural pests, but they can speed up deterioration of wood by transporting moisture into it. Therefore, wood used in construction should not come into direct contact with soil. The ants also become a nuisance when they enter homes in search of food. They require moisture to survive and may be nesting in damp soil outside or under the house, beneath sidewalks, along foundations, or under stones in the yard. Colonies may also be living in moisture-damaged wood around bathtubs, showers, dishwashers, and sinks. The remedy is to remove the decayed wood and replace it with sound material. Remove and replace decayed wood in the basement or crawl spaces of buildings as well, and correct sources of moisture that are responsible for the damage. Chemical control may offer a temporary solution but does not correct the primary problem of wood decay. Ants foraging indoors can be controlled with sugar-based baits (Fig. 2.36).

The large *L. neglectus* infestation in Barcelona mentioned above was controlled with a three-pronged approach: (1) the tree canopy was sprayed with a mixture of cypermethrin and imidacloprid to eliminate the homopterans; (2) the tree trunks were sprayed with cypermethrin to deter foraging ants; and

**Figure 2.36.** *Lasius niger* feeding on bait

(3) foxim was injected into the soil around the outside perimeter of homes, and sugar baits were placed inside (Rey and Espadaler 2004).

Indoor swarmers should be removed with a vacuum sweeper. Management consists of finding the nest and drenching it with a residual insecticide. A sub-slab injection of insecticide may be necessary for nests located beneath foundations (Hedges 1998).

Applications of granular fipronil and commercial baits containing abamectin and hydramethylnon reduced *L. neoniger* mound building on golf courses for at least 60 days (Lopez et al. 2000). Application of pyrethroid sprays as soon as mounds appeared on greens, tees, and adjacent rough suppressed the ants for 4 to 6 weeks; applications later in the spring and summer provided only 2 to 3 weeks' suppression.

### Key to Subgenera of *Lasius*

1 Maxillary palp short; 3-segmented (Fig. 2.37a) . . . . . . . . . . . . . . . . . . . 
. . . . . . . . . . . . . . . . . . . . . . . . . . . . . . . . . . . subgenus *Acanthomyops*
Maxillary palp longer and with 6 segments (Fig. 2.37b) . . . . . . . . . . . 2

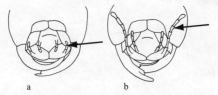

**Figure 2.37.** Ventral view of mouthparts. (a) Subgenus *Acanthomyops* (arrow: three-segmented maxillary palp) (b) *Lasius* spp. (arrow: six-segmented maxillary palp)

2(1) Body shiny black; head with strongly concave occipital margin (Fig. 2.38a) . . . . . . . . . . . . . . . . . . . . . . . . . . . . . . . Subgenus *Dendrolasius*

Body not shiny black; occipital margin of head straight, convex, or slightly concave (Fig. 2.38b) . . . . . . . . . . . . . . . . . . . . . . . . . . . . . . . . . . . . . . . . 3

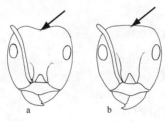

**Figure 2.38.** Face view of head. (a) *L. fuliginosus* (arrow: concave occipital margin) (b) *L. psammophilus* (arrow: straight occipital margin)

3(2) Maxillary palps short, not reaching to middle of ventral surface of head; body yellow to ochreous yellow (Fig. 2.39a) . . . . . . . . . . . . . . . . . . . . . 4

Maxillary palps relatively long, distinctly reaching beyond middle of ventral surface of head; body brownish or grayish black or bicolored with mesosoma lighter than gaster (Fig. 2.39b) . . . . . . . . . . Subgenus *Lasius*

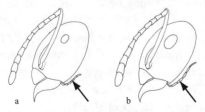

**Figure 2.39.** Profile of head. (a) *L. umbratus* (arrow: short maxillary palp) (b) *L. alienus* (arrow: long maxillary palp)

4(3) Petiolar node viewed from anterior or posterior widest near upper margin; workers polymorphic (Fig. 2.40a) . . . . . . . . . . . . Subgenus *Cautolasius*

Petiolar node viewed from anterior or posterior widest distinctly below upper margin or with parallel sides; workers monomorphic (Fig. 2.40b) . . . . . . . . . . . . . . . . . . . . . . . . . . . . . . . . . . . . . . . Subgenus *Chthonolasius*

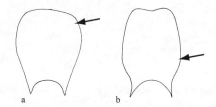

**Figure 2.40.** Posterior view of node. (a) *L. flavus* (arrow: wide apex) (b) *L. subumbratus* (arrow: width greater below the apex)

## Rover Ants

*Brachymyrmex* species

### IDENTIFYING CHARACTERISTICS

Rover ants are small (1.5–2 mm long), nondescript ants; workers are monomorphic (Snelling and George 1979). The antenna of workers has nine segments and lacks a club. The body color varies from yellow to black depending on the species (Vail et al. 1994).

### DISTRIBUTION

Most species of this small New World genus are tropical (Snelling and George 1979); only *B. depilis* has been described as occurring naturally in the United States. Its range is Nova Scotia to British Columbia and south to central Mexico (Snelling and George 1979). Nevertheless, *Brachymyrmex* is in "a state of taxonomic chaos," and there may be as many as six to ten North American species (Fisher and Cover 2007). Introduced rover ant species such as *B. patagonicus* and *B. obscurior* are minor household pests in various regions of the country, particularly the Southeast. For example, a year-long structural pest control survey in Florida reported fifteen cases involving these two species (Klotz et al. 1995). The first reported case of *B. obscurior* in Georgia was of an infestation in a laboratory (Ipser et al. 2005).

### BIOLOGY AND HABITS

*Brachymyrmex* colonies are small and monogynous (Hedges 1998). Nests are in soil or rotting wood (Vail et al. 1994), and around homes may be found in potted plants and under objects in gardens (Dash et al. 2005). Large numbers of rover ants are often noticed after suppression of imported fire ants (Dash et al. 2005).

Mating flights occur from early to midsummer in Louisiana (Dash et al. 2005). In Florida, the alates are attracted to lights and are often found in bathrooms (Vail et al. 1994). In the Florida survey mentioned above, alates were collected in seven of the fifteen cases; in two of these, hundreds of alates were collected in swimming pools.

Foraging workers tend subterranean homopterans on plant roots and harvest their honeydew. Occasionally, workers invade homes in large numbers, infesting kitchens, bathrooms, light sockets, electrical outlets, and boxes associated with water heaters (Vail et al. 1994; Dash et al. 2005).

## CONTROL

Alates within a structure can be removed with a vacuum sweeper (Hedges 1998). Indoor infestations are often associated with excessive moisture or fungal decay; thus, correcting moisture problems and removing damaged wood are important control measures (Hedges 1998). Nests located in void spaces may be treated by drilling and injecting an aerosol or dust over the colony. Sub-slab infestations may need to be treated by drilling and injecting a residual insecticide. Outdoor nests should also be located and drenched with a residual insecticide (Hedges 1998). Some PMPs apply an outside perimeter spray of fipronil and an interior crack and crevice treatment with chlorfenapyr.

### Key to Species of *Brachymyrmex*

1 Body yellow . . . . . . . . . . . . . . . . . . . . . . . . . . . . . . . . . . . . . . . *B. depilis*
  Body dark brown to gray . . . . . . . . . . . . . . . . . . . . . . . . . . . . . . . . . 2

2 (1) Dense pubescence and erect hairs on gaster and pronotum . . . . *B. obscurior*
  Two pairs of bristles on dorsum of mesosoma, no dense pubescence on gaster . . . . . . . . . . . . . . . . . . . . . . . . . . . . . . . . . . . . . . . . *B. patagonicus*

## Long-Legged Ant

*Anoplolepis gracilipes* (Plate 2a)

### IDENTIFYING CHARACTERISTICS

Workers are monomorphic, about 4 mm long, and move very quickly on their very long legs. Their antennae are eleven-segmented and also very long. The head is small with large eyes, and the body is yellow-brown with a darker abdomen (Wetterer 2005).

DISTRIBUTION

The long-legged ant is a tramp species that originated in Africa or tropical Asia (McGlynn 1999) and has been distributed to many tropical countries (Haines et al. 1994). In Hawaii, this ant is typically found in lowland rocky areas with moderate rainfall and is one of the most commonly encountered species in buildings (Reimer et al. 1990). The older literature refers to this species as *A. longipes,* but the oldest junior synonym, *A. gracilipes,* took precedence over that name.

BIOLOGY AND HABITS

Also known as the crazy ant because of its quick, erratic movements (Haines et al. 1994), *A. gracilipes* is among the most widely distributed, abundant, and damaging invasive ant species in the world (Holway et al. 2002). Although the species has been poorly studied (Holway et al. 2002), colonies are reported to be unicolonial, polygynous, and polydomous.

Research on the species's biology has been conducted on the Seychelles Islands in the Indian Ocean, where the long-legged ant is an agricultural and household pest and may have a negative impact on the islands' biodiversity (Haines et al. 1994). Nests are typically constructed on the ground in crevices and under stones and coconut husks, but arboreal nests—for example, in coconut palms—are also known. Colonies range in size from 2,500 to 36,000 workers and have 1 to 320 queens per nest; supercolonies may have 5 to 10 million ants per hectare. The ants are active year-round with no evident seasonality. Reproductives are present throughout the year. The primary mode of colony multiplication appears to be budding; no mating flights have been observed. Long-legged ants are omnivorous, and workers forage both day and night for dead and dying invertebrates and, especially, honeydew. They also feed on juices from fruit, plant exudates, and occasionally on vertebrates.

CONTROL

The control program in the Seychelles focused on the development of baits for areawide management and contact sprays for urban dwellings (Haines et al. 1994). The most effective baits contain sugar, salt, and yeast extract as the attractants and phagostimulants. The most effective spray for indoor control is bendiocarb (0.15% AI).

## *Plagiolepis* **Species (Plate 2b)**

### IDENTIFYING CHARACTERISTICS

The workers are small, ranging in size from 1.2 to 2 mm. The body is compact and arc shaped when viewed from the side, and the propodeum is smooth and rounded. The antenna has eleven segments, and individuals of most species are yellow to light brown. They are sometimes confused with ghost ants, which infest similar places such as tropical greenhouses (Andersen 1991; Shattuck and Barnett 2001).

### DISTRIBUTION

Fifty-six species of *Plagiolepis* have been described (Shattuck and Barnett 2001, Bolton et al. 2006); most are found in the tropics and subtropics. They are native to southern Europe and Asia, Africa, and Australia. Species introduced into central Europe on tropical plants have spread to many tropical greenhouses. *Plagiolepis alluaudi* is among the most commonly encountered pest ants in buildings in Hawaii (Reimer et al. 1990) and has also been collected on the Channel Islands off the California coast (McGlynn 1999).

### BIOLOGY AND HABITS

*Plagiolepis* species nest in the soil under logs, rotten wood, and stones, and in cavities beneath bark. The nests are small, usually not exceeding 20 mm in diameter. Nests in greenhouses in Europe are typically found between roots and inside bark crevices and flowerpots (Fig. 2.41). These species usually require high humidity; however, colonies may survive in a dry room if water is available.

The workers forage for nectar and honeydew, although their small size makes tending aphids more difficult than it is for larger ants such as *Lasius* species (Seifert 1996). One of the authors (RP) observed them in a greenhouse feeding on honey, sugar water, and pieces of overripe bananas that were provided as food for tropical butterflies. They were primarily active at night, with some activity during the day. Others have observed these ants in tropical greenhouses collecting honeydew from homopterans (Sellenschlo 2002b).

### CONTROL

To prevent infestations in tropical greenhouses, new plants should be thoroughly inspected, treated, and quarantined for several weeks. Sugar-based

**Figure 2.41.** Nest of *Plagiolepis* spp. in the soil of a flowerpot

baits and insecticidal sprays and dusts can be used for control, although the latter must be carefully applied so as not to injure the plants. Additional measures include reducing aphid populations and eliminating decaying fruit, which serve as food sources for the ants.

# CHAPTER THREE

# Dolichoderinae

## Subfamily Characteristics

In general, members of this subfamily resemble formicines, but the anus is ventral rather than terminal and transverse linear rather than circular. Dolichoderines have a one-segmented petiole, and the sting is vestigial or absent (Bolton 1994). Instead, they use chemical defenses, releasing strong and repugnant volatile odors from the anal gland when threatened (Fisher and Cover 2007). They are distributed worldwide and are typically soil-nesting omnivores.

SCIENTIFIC AND COMMON NAMES

*Forelius pruinosus* (Roger, 1863)
*Dorymyrmex:* Pyramid ants
    *D. bicolor* W.M. Wheeler, 1906
    *D. insanus* (Buckley, 1866)
*Linepithema humile* (Mayr, 1868): Argentine ant
*Liometopum*
    *L. occidentale* Emery, 1895: Velvety tree ant
    *L. apiculatum* Mayr, 1870
    *L. luctuosum* W.M. Wheeler, 1905
*Tapinoma*
    *T. sessile* (Say, 1836): Odorous house ant
    *T. melanocephalum* (Fabricius, 1793): Ghost ant
*Technomyrmex albipes* (F. Smith, 1861): White-footed ant

## Key to Genera of Dolichoderinae

1  Node on pedicel flattened when viewed in profile (Fig. 3.1a). . . . . . . . . 2
   Node on pedicel vertical when viewed in profile (Fig. 3.1b) . . . . . . . . . 4

**Figure 3.1.** Mesosoma and gaster. (a) *Tapinoma sessile* (arrow: flattened node) (b) *Linepithema humile* (arrow: vertical node)

2(1)  White gaster with dark mesosoma and head. . .*Tapinoma melanocephalum*
      Gaster, mesosoma, and head uniformly dark . . . . . . . . . . . . . . . . . . . . 3

3(2)  Dark body and white or light yellow tarsi . . . . . . *Technomyrmex albipes*
      Body, legs, and tarsi brown . . . . . . . . . . . . . . . . . . . . . *Tapinoma sessile*

4(1)  Single, conical projection on dorsum of epinotum (Fig. 3.2a)
      . . . . . . . . . . . . . . . . . . . . . . . . . . . . . . . . . . . . . . . . . . *Dorymyrmex* spp.
      No projections on dorsum of epinotum (Fig. 3.2b) . . . . . . . . . . . . . . . 5

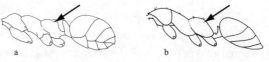

**Figure 3.2.** Mesosoma and gaster. (a) *Dorymyrmex insanus* (arrow: conical projection) (b) *Forelius pruinosus* (arrow: no projection)

5(4)  Mesosomal dorsal profile evenly or nearly evenly rounded (Fig. 3.3) . . . .
      . . . . . . . . . . . . . . . . . . . . . . . . . . . . . . . . . . . . . . . . . . *Liometopum* spp.
      Mesosomal dorsal profile not evenly rounded (Fig. 3.1b). . . . . . . . . . . 6

**Figure 3.3.** Mesosoma and gaster of *Liometopum* spp. (arrow: smooth thoracic dorsum)

6(5)  Mandible with 2 large apical teeth and fine denticles; mesosoma without
      erect hairs; epinotum in profile short, approximately twice as high as long
      (Fig. 3.1b); body uniform brown . . . . . . . . . . . . . . . *Linepithema humile*
      Mandible with small or large teeth but no fine denticles; mesosoma with a
      few erect hairs on pronotum; epinotum in profile not twice as high as long
      (Fig. 3.2b); body color variable. . . . . . . . . . . . . . . . *Forelius pruinosus*

# Argentine Ant

*Linepithema humile* (Plate 2c)

## IDENTIFYING CHARACTERISTICS

Workers are 2.2 to 2.6 mm long with a one-segmented petiole and a twelve-segmented antenna without a club (Smith 1965). The body color varies from light to dark brown with somewhat lighter legs (Vega and Rust 2001). The mandibles are yellowish and dentate. Argentine ants cannot sting but emit a musty odor when crushed (Smith 1965). Queens are brown and 4 to 6 mm long. They are not mere egg-layers as are queens of most other species but also perform duties such as foraging and feeding and grooming the colony's young (Vega and Rust 2001). Males are dark brown and 2.8 to 3 mm long.

## DISTRIBUTION

Argentine ants originated in the Paraná River basin of subtropical South America (Wild 2004) but are now found worldwide in areas with a mild temperate Mediterranean climate (Holway et al. 2002). They probably reached the United States in the late 1800s on ships transporting coffee from Brazil to New Orleans (Newell and Barber 1913). They are now established throughout the southeastern United States and are also found in Maryland, Illinois, Indiana, Missouri, Oklahoma, Arizona, California, Oregon, Washington, and Hawaii (Hedges 1998; Bolton et al. 2006). Genetic studies indicate that the Argentine ant population in California was introduced from the southeastern United States, thereby undergoing a double genetic bottleneck (Buczkowski et al. 2004). As a consequence, the California population has less genetic diversity and exhibits less intercolony aggression than the southeastern population.

Argentine ants have also been transported to Europe, Australia, South Africa, Central America, the Mediterranean, and Caribbean and Atlantic islands (McGlynn 1999). There are two enormous supercolonies in southern Europe, the largest one stretching more than 6000 km (3728 mi) along the coast from Portugal to northern Italy (Giraud et al. 2002). Large colonies are found in urban areas in the Lower Rhine district (Germany) and in the Netherlands.

## BIOLOGY AND HABITS

The Argentine ant is both an invasive and a tramp species (McGlynn 1999). It thrives in disturbed habitats with abundant moisture and low ant diversity (Majer 1994). As an invasive species, the Argentine ant becomes established in an area and then spreads into the surrounding environment, aggressively displacing other native ants (McGlynn 1999). The colonies lack borders and can extend over entire habitats. Indeed, their unicoloniality is a key attribute of Argentine ants' ecological success (Giraud et al. 2002). In their native South America, however, colonies coexist with other ant species and maintain colony borders, which they defend against neighboring Argentine ant colonies (Suarez et al. 1999). The loss of intraspecific aggression in introduced populations gives Argentine ants a competitive edge over other less populous species (Tsutsui and Suarez 2003).

In common with other tramp species, Argentine ants are adaptable and easily transferred by commerce (McGlynn 1999). In addition, they are polygynous, unicolonial, and achieve colony multiplication by budding, which gives colonies tremendous capacity for growth and expansion (Majer 1994; Passera 1994). The colony structure is fluid, with workers, brood, and food moving between nests depending on the distribution of resources (Holway and Case 2000). Argentine ant populations can reach astronomical proportions. In a citrus grove in San Diego County, California, for example, 50,000 to 600,000 ants ascended each tree daily to tend homopterans (Markin 1967); and in a residential area in Southern California, an estimated 176,000 to 538,000 ants visited each home over a 24-hour period (Reierson et al. 1998).

The typical queen-to-worker ratio is 15:1000 (Aron 2001), but this number is reduced to 1:1000 after the workers execute queens at the beginning of the reproductive season in the spring (Keller et al. 1989). Queen execution is believed to trigger the production of new reproductives (Keller et al. 1989) and to increase genetic relatedness, thereby promoting colony cohesion (Tsutsui and Suarez 2003).

Virgin queens do not disperse on nuptial flights. They mate within the nest, although without a significant level of inbreeding (Krieger and Keller 2000), possibly because males have mating flights, and queens and sexual brood are mobile (Kaufmann et al. 1992). New colonies are initiated by budding from the parent colony. Small propagules consisting of workers and brood either with or without queens can grow and reproduce quickly (Aron 2001). Males either mate with virgin queens in their own nest or leave on mating flights to mate with females in other nests.

The eggs, larvae, and pupae remain hidden within the underground nest and are seen only when nests are disturbed or when workers are carrying them to a different location (Fig. 3.4). The microscopic eggs are white and approximately 0.3 mm long (Newell and Barber 1913). Queens lay eggs throughout the year, but most eggs are produced in spring and summer. The incubation pe-

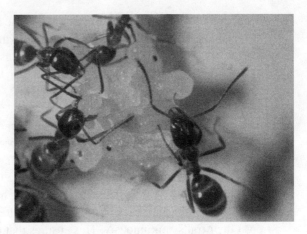

**Figure 3.4.** Argentine ant (*Linepithema humile*) workers and brood

riod varies from 12 to 55 days, depending on the temperature, and averages 28 days (Newell and Barber 1913). The length of time spent in the larval and pupal stages also varies depending on temperature. After four larval instars the larvae molt into pupae, which look like adults with their legs and antennae held tight against the body. The pupae are initially white but turn darker as they mature. In the final molt, an adult ant emerges from the pupa. The time needed to complete the cycle from egg to adult ranges from 33 to 141 days, with an average of 74 days (Newell and Barber 1913).

The colony cycle is seasonal. In southern California, for example, worker production starts in mid-March and ends in October (Markin 1970a). Colonies grow throughout the spring and continue growing until about mid-October to November, when a massive die-off occurs (Reierson et al. 2001). Reproductives overwinter with workers, and in the late winter and early spring begin laying eggs that will develop into new alates and workers.

In winter, Argentine ants often nest out in the open, and small piles of excavated soil reveal their nest sites. During summer, they move into shade to avoid direct sunlight on the nest (Markin 1967). In adverse conditions or as winter approaches, colonies sometimes merge and form supercolonies (Markin 1968). With the coming of spring, these supercolonies break up into smaller colonies that disperse to find new nest sites (Barber 1920).

Nests typically are shallow (20 cm, 0.66 ft), but in dry soils they can be as deep as 60 cm (1.97 ft) (Markin 1967). In urban areas, outside nests are located beneath boards, stones, and concrete, and within decaying plant matter and mulch. Nests are often found at the base of plants or trees infested with homopterans. Indoor infestations tend to occur in kitchens and bathrooms, where water and food are available. Indeed, one of the authors (RP) was able to capture a small indoor colony using a Petri dish that contained only moistened plaster; the entire colony, including the queens and brood, moved into this makeshift nest. Indoor nests are often located in voids below kitchen and bathroom cabinets, around and behind dishwashers, and under stairwells (Gulmahamad 1996). Potted houseplants, bath traps, and subslab areas sometimes harbor nests as well (Gulmahamad 1996). Homeowners sometimes complain of finding huge numbers of dead ants in unexpected places, such as in a bathtub or on the garage floor. These corpse piles are probably debris from large colonies doing their "spring cleaning" after the fall and winter die-off.

Argentine ants forage systematically. Unlike many ants, which deposit odor trails only on the way back to the nest from a food source, Argentine ants deposit their trails continuously. This ensures that they cover new ground when searching for food and are not revisiting the same locations (Deneubourg et al. 1990). When they find food, the ants reinforce the trail to recruit additional ants to collect the resource. Once trails are established individual ants show a high degree of site fidelity to a specific resource (Fernandes and Rust 2003). Argentine ants are often seen trailing along sidewalks, driveways, and other structural edges that provide guidelines for foraging ants. Trails may consist

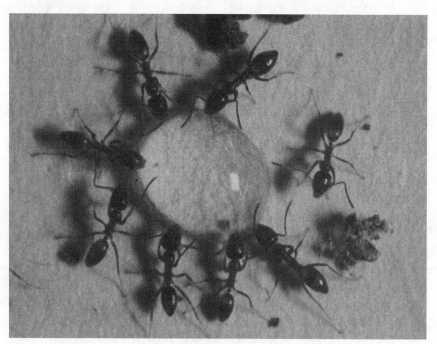

**Figure 3.5.** Argentine ant (*Linepithema humile*) workers around liquid bait

of thousands of ants traveling to and from nests. Vega and Rust (2003) fed Argentine ants dyed sucrose water and found a large percentage of dyed ants 60 m (197 ft) from the food source.

Approximately 99% of the food brought into an Argentine ant nest is in liquid form (Markin 1970b) (Fig. 3.5). Although the diet changes over the year, the main dietary staple is honeydew (Markin 1970a), which consists of sugars, amino acids, lipids, vitamins, and minerals. The colony's acceptance of sucrose water or honey decreased from December through February (Rust et al. 2000), and the amount of prey brought into the nest fluctuated from one ant per thousand carrying prey in November to ten ants per thousand in May (Markin 1970b).

## Control

Seasonal patterns of Argentine ant infestations in the urban environment vary by region. For example, in Berkeley, California, an initial invasion of buildings in August coincides with the decline of homopteran populations, and a second peak occurs in fall when the first substantial rains inundate colonies (Art Slater, pers. comm., 2005). Invasions in southern California usually begin in early June and end by late summer, with indoor infestations being associated with increasing temperature and aridity, reduced homopteran numbers, and rainfall (Rust et al. 1996). Ideally, control measures for Argentine ants should be initiated early in the season when colonies are undergoing their most rapid growth and development. Stunting colony growth early will result in decreased numbers later in the season.

Perimeter treatments with residual insecticides or toxic baits are currently the primary methods of controlling Argentine ants. Various pyrethroid sprays and granular formulations, for example, are in common use to create barriers around homes. A relatively new termiticide, Termidor®, is also registered for use against Argentine ants. The active ingredient, fipronil, is slow-acting, non-repellent, and readily transferred from one ant to another, acting more like bait than a barrier (Vail et al. 2003). Consult the label for the correct application rate. Some product labels permit barriers of up to 3.05 m (10 ft) from the foundation. Termidor® sprays can be applied only twice per year and are restricted to a 0.3-m (0.98-ft) band on the ground at the base of the foundation and a 0.3-m band up the foundation. Approximately 5.7 l (1.5 gal) of finished spray may be applied per 92.9 m$^2$ (1000 ft$^2$) (Potter and Hillery 2003).

Depending on the insecticide and formulation, barriers have different degrees of contact activity and repellency (Knight and Rust 1990b). Some pyrethroid formulations are highly toxic to ants and often create additional problems when ants indoors are cut off from their colony outdoors. The trapped ants may increase their activity inside searching for an escape route, thus increasing their visibility. Synthetic pyrethroids such as bifenthrin (Talstar®) suppress recruitment, giving the appearance of being highly repellent (Richman and Hooper-Bui 2003). Applied at the maximum label rate, a pyrethroid can provide 8 to 10 weeks of control (Klotz et al. 2002).

Creating an effective barrier against Argentine ants is difficult because any small gap will provide an opening. Other factors that may reduce efficacy include chemical degradation, irrigation, dense groundcover, mulch, high temperature, substrate alkalinity, and direct sunlight (Rust and Knight 1990; Rust et al. 1996). To be most effective, a residual insecticide should be applied thoroughly, not only as a band around the foundation but also where the ants trail, such as along the edges of sidewalks and steppingstones; at the base of trees, potted plants, and garbage cans; and on nests (Rust et al. 1996). One drawback to a perimeter treatment is its broad-spectrum effects; it may kill beneficial arthropods, resulting in secondary pest outbreaks (Smith et al. 1996).

In contrast, toxic baits are more target-specific and environment-friendly, particularly when delivered in bait stations and not broadcast. Populations of Argentine ants are particularly vulnerable to toxic baits because ants move constantly between nests. More than 50% of the worker population of one colony was exchanged among neighboring nests in a 5-day period (Markin 1968). Thus, bait consumed by ants from one nest will gradually spread into surrounding nests. Toxic baits are particularly effective when applied early in the seasonal life cycle of Argentine ants, when the colony's demand for food is high but food availability is relatively low. Protein-based baits are most attractive at this time of the year (Rust et al. 2000), probably because egg production is higher and the number of developing brood is increasing.

Unfortunately, most of the commercial baits available are not attractive to Argentine ants, and those that are kill the ants before the bait can be dispersed through the colony (Rust et al. 2002). Under current development, however, are some sugar-based liquids containing new active ingredients effective at ultra-low concentrations (Rust et al. 2002). These sucrose water baits capitalize on the Argentine ant's preference for honeydew (Markin 1970c) and its digestive tract and foraging behavior, which are specialized for a liquid diet (Hölldobler and Wilson 1990), but the ant's foraging behavior can create problems. Argentine ants typically forage over distances of 60 m or more. In a residential setting, this may encompass several homes and will require areawide control. Another challenge is how to redirect foragers from an established trail to a toxic bait (Fernandes and Rust 2003). One possibility is a pheromone-enhanced guideline, such as a string treated with Z9-16:Ald (synthetic trail pheromone), to draw ants off a trail and into a bait station (Greenberg and Klotz 2000).

For a comprehensive historical review of control strategies for Argentine ants, both chemical and nonchemical, see Vega and Rust 2001. Some promising new research on the toxicity and repellency of aromatic cedar mulch to Argentine ants suggests its use in urban integrated pest management (IPM) programs (Meissner and Silverman 2001, 2003). Landscape mulches such as pine straw provide ideal nesting sites for ants, while cedar mulch not only repels Argentine ants but is lethal when they travel across it. Another substrate that inhibits foraging is fine sand (i.e., fine enough to pass through 200 mesh) (Rust et al. 2003).

## Odorous House Ant

*Tapinoma sessile* (Plate 2d)

### Identifying Characteristics

Workers are monomorphic but vary in length from 2.4 to 3.3 mm. The petiole has one segment, and its node is flattened and is concealed by the base of the gaster when viewed from above or from the side (Fig. 3.6). These ants are uniform brown to black, and when crushed emit an odor reminiscent of rotten coconut (Ebeling 1975). *Tapinoma sessile* shows great diversity in DNA over its entire range and may well represent several species (L. Davis Jr., pers. comm., 2007).

**Figure 3.6.** Odorous house ant (*Tapinoma sessile*) worker with larva

DISTRIBUTION

The odorous house ant is found throughout the United States and southern Canada at altitudes from sea level to above 3352.8 m (11,000 ft) (Wheeler and Wheeler 1986), and is thought to be undergoing an extensive range expansion (Scharf et al. 2004). Indeed, it may have the widest distribution and greatest ecological tolerance of all the North American ant species (Fisher and Cover 2007). It is common in parts of Tennessee, Arkansas, Mississippi, California, and the Pacific Northwest (Hedges 1997), and is becoming a serious pest in some midwestern and mid-Atlantic states, particularly Kentucky, Virginia, and New Jersey (Hedges 2002).

BIOLOGY AND HABITS

Colony sizes ranges from 2000 to 10,000 ants (Snelling and George 1979), and each colony contains numerous reproductive females (Smith 1928). There are anecdotal reports of much larger colonies (>100,000) with foraging trails more than 30.5 m (100 ft) long.

The winged male and female reproductives mate either in the nest or after a nuptial flight (Smith 1965). Female alates have been collected during May in desert localities (Snelling and George 1979). Colonies are established by newly mated queens or by budding from preexisting colonies. Colony budding and migration may be precipitated by chemical or mechanical disturbance (Barbani and Fell 2002) or in response to rain (Vail et al. 2003).

Ants from different colonies are not aggressive toward one another (Smith 1965). Indeed, with the presence of many queens in many nest sites, colony boundaries are difficult to determine. Odorous house ants are submissive or subordinate to other species of ants, such as honey and acrobat ants, which frequently displace them at food sources (Barbani and Fell 2002). Some researchers believe that the elimination of other ants with perimeter treatments has provided a competitive release for odorous house ants that is contributing to their increase (Scharf et al. 2004). The increasing use of landscape mulch, which provides prime nesting material for odorous house ants, and higher temperatures due to global warming may also be contributing to their growing prevalence in urban situations (Scharf et al. 2004).

The duration of each developmental stage varies depending on the season; ranges are 11 to 26 days for eggs, 13 to 29 days for larvae, 2 to 3 days for prepupae, and 8 to 25 days for pupae (Smith 1928).

The odorous house ant is an opportunistic nester. Nests in soil are usually shallow and tend to be located beneath objects such as boards, stones, or other debris (Smith 1965). Stacked siding, lumber, firewood, bricks, and cardboard are also favorite nesting sites (Hedges 1998). Although the nests of odorous house ants that invade buildings are often located outside, these ants nest inside structures as well. Inside nests are usually associated with moisture such as that found within wall voids near pipes and heaters, bath traps, wood damaged by termites, and beneath toilets (Vail et al. 2003). Odorous house ants are also common pests in apiaries, nesting beneath the top and inner covers of beehives (Hedges 1997). When households move from one area to another, the ants often go along. In eastern Washington, for example, these ants are commonly found around mobile home parks.

The workers forage both day and night. Honeydew is among their favorite foods (Snelling and George 1979), and they are attracted to sugars year-round (Barbani and Fell 2002). They also feed on live and dead insects (Smith 1965) and are attracted to cat food. Odorous house ants can be active throughout the year and have been seen foraging outdoors at temperatures as low as 10 °C (50 °F) (Smith 1928).

CONTROL

A recent study combined bait with a residual spray treatment to control odorous house ants in and around homes (Vail et al. 2003). Bait stations containing liquid bait with 1.3% borax were placed outside the boundary of a nonrepellent perimeter spray of 0.06% fipronil applied to the foundation 0.3 m up and out. This treatment was compared with a nonrepellent perimeter treatment only and a bait treatment only. Of the three treatments, the bait—residual spray combination was the most effective because it quickly reduced both indoor and outdoor populations of ants. Baits containing boric acid and imidacloprid were effective in reducing the number of odorous house ants around structures for 8 weeks (Higgins et al. 2002). This species is much more susceptible to baiting than Argentine ants, and proper identification is thus extremely important.

## Ghost Ant

*Tapinoma melanocephalum* (Plate 2e)

IDENTIFYING CHARACTERISTICS

Workers are 1.3 to 1.5 mm long and have a one-segmented petiole. The petiolar node is flattened and hidden from view dorsally and laterally by the base of the gaster. The head, antennal base, and thoracic dorsum are dark brown (this species is also known as the black-headed ant); the gaster, legs, and distal antennal segments are light yellow, almost translucent. Queens are light to medium brown and about 2.6 mm long.

## DISTRIBUTION

Ghost ants have been widely disseminated by commerce from their African or Asian point of origin (Smith 1965). They travel readily in plants or even luggage. In the United States, they are common household pests in central and south Florida (Hedges 1992) and are also found in Hawaii, southeast Texas, and sporadically in California and the Pacific Northwest (Hedges 1997). In addition to North America, ghost ants have reached Europe, Central and South America, Southeast Asia, Australia, and the Arabian Peninsula (McGlynn 1999). In Europe they are found only indoors in continuously heated environments such as apartment houses, shops, tropical greenhouses and zoological gardens. In the Netherlands and United Kingdom, for example, long-standing infestations of at least seven years were found in apartment complexes, and in Germany, infestations have been reported in a flight room for tropical butterflies, a food preparation area of an aquarium, and in apartment complexes with central heating. Several infestations have been reported in Switzerland.

## BIOLOGY AND HABITS

Colonies of ghost ants vary in size (100–1000 individuals) and have numerous queens (Harada 1990). New colonies are established by budding, and workers following odor trails move freely between nests (Harada 1990). It is not known if ghost ants have mating flights.

Ghost ants require high humidity. They nest in soil, rotten wood, under bark, and in plant cavities and detritus. In Florida, nests are commonly found at the base of palm fronds in decaying organic matter (Vail et al. 1994). In households, they nest in cracks and crevices (Fig. 3.7), potted plants, breadboxes, shower curtain rods, behind baseboards, between cabinets, and even inside clothes irons (Hedges 1992) and between books. The ants enter buildings on utility lines and through cracks around windows, doors, and soffits. Outside nests tend to be in soil next to foundations, porches, shrubs, and trees. Hollows in pool enclosures are also a favorite nesting site (Hedges 1997). For successful development, the brood requires a temperature above 25 °C and high humidity.

Indoors, ghost ants' activity is typically concentrated in the kitchen or bath-

**Figure 3.7.** Ghost ants (*Tapinoma melanocephalum*) trailing along a crack in concrete

room near sources of water, although Hedges (1997) found them trailing from room to room under the edge of carpeting. In kitchens, they prefer to forage on sweets but will also feed on grease. The workers can penetrate packaging through even very tiny cracks. They tend homopterans for honeydew and feed on both live and dead insects.

In some areas of the world, ghost ants are major structural pests. In Valley Province, Colombia, for example, they were found in 51% of hospitals and 29% of residences; and in residences and industrial kitchens in São Paulo, Brazil, they were mechanical vectors of pathogenic bacteria (Ulloa-Chacon and Jaramillo 2003; Zarzuela et al. 2005). In greenhouses, ghost ants may become a problem by eating beneficial insects introduced for biological control.

## CONTROL

Infestations are easily overlooked because ghost ants are so small. Only a thorough inspection with special attention to food storage areas with high temperature and humidity is likely to reveal them.

Laboratory research suggests that sweet liquid baits may be an effective control measure for ghost ants. Colonies were eliminated with boric acid (0.5% and 1%) or fipronil (0.05%) in 10% sucrose water (Ulloa-Chacon and Jaramillo 2003). In Florida, where the ghost ant is a major household pest, the

Cooperative Extension Service recommends barrier sprays to prevent foraging ghost ants from entering homes (Nickerson and Bloomcamp 2003).

Utility lines should be sealed at their point of entry to structures. Other preventative measures include good sanitation, storing food in tight containers, removing plants that attract homopterans, and eliminating sources of moisture such as condensation and leaks (Nickerson and Bloomcamp 2003). Reducing humidity is particularly important in infested apartment buildings, hotels, and restaurants.

## White-Footed Ant

*Technomyrmex albipes* (Plate 2f )

### IDENTIFYING CHARACTERISTICS

Workers are 2.5 to 3 mm long and have a one-segmented petiole with a flattened node that is concealed by the base of the gaster when viewed from above or from the side. They are black to brownish black with yellowish white tarsi (feet) and do not bite or sting (Warner et al. 2004). *Technomyrmex albipes* workers resemble *Tapinoma* workers but have five visible gastral tergites rather than four (Vail et al. 1994).

### DISTRIBUTION

The white-footed ant is a tramp species native to the Pacific and Indo-Australian regions (Wilson and Taylor 1967) that has spread on nursery stock and other commercial shipments. It is widely distributed throughout central and south Florida, and small infestations have been reported in North Carolina, South Carolina, Georgia, and Louisiana (Warner and Scheffrahn 2004a). Pending reclassification, the U.S. species may in fact be a closely related species, *T. difficilis* (Fisher and Cover 2007).

White-footed ants have spread to Guam and the Hawaiian Islands, and have reached pest status on Oahu. Several established colonies have been reported in San Francisco hothouses (Ward 2005) and in the Seattle Zoo. These ants have also reached southern Africa, India, China, Europe, and Saudi Arabia (McGlynn 1999). Three infestations were reported in Switzerland in 2005 (Umwelt-und Gesundheitsschutz Zuerich 2004), and a large infestation was found in a tropical greenhouse in Bonn, Germany, where the ants were interfering with a biological control program to control aphids.

## Biology and Habits

Colonies of white-footed ants are decentralized and comprise many satellite nests extended over a large territory that regularly exchange workers, brood, and food. Colonies can reach a tremendous size with several million workers and numerous reproductive females called intercastes (Yamauchi et al. 1991).

Mating flights occur between July and August in southern Florida. Inseminated queens found new colonies but are eventually replaced by intercastes (Warner et al. 2004). Colony multiplication also occurs by budding, in which intercastes leave the parent colony along with some nestmates and brood to establish a new nest (Warner et al. 2004).

Nests tend to be arboreal but have also been located in the soil at the base of trees and on trees under loose bark (Hedges 1997). Nests are occasionally found inside structures as well, in wall voids and attics (Warner et al. 2004). White-footed ants are more frequently pests in houses than in commercial buildings (Hedges 1997) and can short-circuit air conditioners (Mangold 1996).

White-footed ants are particularly attracted to sweets, and they feed on nectar and tend homopterans for honeydew. They also eat dead insects and other sources of protein (Warner et al. 2004). Their transfer of nutrients is unusual because it occurs through trophic eggs rather than trophallaxis (Yamauchi et al. 1991).

## Control

White-footed ants are considered nuisance pests when they occur in structures and landscapes in large numbers (Warner and Scheffrahn 2004a). They are among the most difficult ants to control because their colonies are so large (Warner et al. 2004). Researchers at the University of Florida have been at the forefront in developing control strategies. Their recommendations include a comprehensive baiting program using sweet liquid borate baits placed both inside and outside wherever there are nests or trailing activity, including the landscape around infested vegetation. Sucrose water at concentrations of 25 to 40% makes an adequate attractant, and up to 7% disodium octaborate tetrahydrate is not repellent (Warner and Scheffrahn 2004a). High concentrations of borates (ca. 5%) are recommended because delayed toxicity is not a limiting factor as it would be if the ants shared food by trophallaxis. In addition to borates, ultra-low concentrations of thiamethoxam and imidacloprid are effective in sweet liquid baits (Warner and Scheffrahn 2005).

Residual insecticides can be applied to nests and trails as a follow-up to baiting, but they are ineffective as stand-alone treatments (Warner and Scheffrahn 2005). Creating vegetation-free zones around structures to eliminate potential access points for the ants is a critical nonchemical control measure.

## Velvety Tree Ants

*Liometopum* species (Plate 3a)

### IDENTIFYING CHARACTERISTICS

Workers of this small genus of arboreal ants (Snelling and George 1979) are moderately polymorphic and range from small to medium in size. The petiole has one segment with a vertical node that is somewhat hidden by the gaster. The abdomen is covered with fine hairs that give a velvety appearance. They are often mistaken for carpenter ants, particularly *Camponotus clarithorax*, which is similar in size and color (Gulmahamd 1995). Velvety tree ants may be black with a red mesosoma, uniform brown, or black, and when crushed they emit a rotten coconut-like odor very similar to that produced by the odorous house ant. They excavate wood and insulation, producing a fine-textured frass, and like carpenter ants their mesosoma has a smooth, even profile (Hedlund 2003). The cloacal opening is slitlike, however, rather than circular and fringed with hairs like that of carpenter ants.

### DISTRIBUTION

Three species of velvety tree ants are found in the western and southwestern United States. The "velvety tree ant" (*Liometopum occidentale*) occurs along the West Coast from Washington to Baja California at altitudes from sea level to 1463 m (4800 ft) (Snelling and George 1979), and is commonly found on oak, elm, cottonwood, pine, sycamore, and alder (Gulmahamad 1995). The closely related *L. luctuosum* ranges from Wyoming to western Texas to Nevada and California (Gulmahamad 1995) and the Pacific Northwest (Merickel and Clark 1994). It nests and forages in conifers, often at higher elevations than its congeners; in Nevada, for example, this species has been found at altitudes of 1372 to 2469 m (4500–8100 ft) (Wheeler and Wheeler 1986). Regional common names for *L. luctuosum* include black velvety, pine tree, and silky carpenter ant. The other montane species, *L. apiculatum,* ranges from Colorado through Arizona, New Mexico, and Texas (Gulmahamad 1995) in foothills areas up to 2133 m (7000 ft); it is often associated with live oak (Creighton 1950).

**Figure 3.8.** Wall damage by velvety tree ants (*Liometopum luctuosum*)

## BIOLOGY AND HABITS

In their natural habitat, velvety tree ants typically nest in soil or duff in or under decaying logs, under rocks, and in the crevices or hollows of trees (Gulmahamad 1995). They often infest structures and excavate wood (Fig. 3.8) or insulation to construct their nests or temporary resting places. Infestations resemble those of carpenter ants, but the excavated material has a finer texture. The nest chambers are composed of "carton" (Wheeler 1905), which the ants make by mixing pieces of plant material and soil and adding an oral secretion that cements it together, forming a mass of trabeculae. Brood chambers are maintained within this matrix of paperlike material.

When they nest inside structures, velvety tree ants can produce considerable amounts of frass consisting of chewed wood and insulation. Foragers often congregate in hollows within insulation or wood, forming temporary "resting places," or bivouacs. Attempts to eradicate nests indoors are often unsuccessful because the bivouacs, which are vacated or moved as foraging territories change, are mistaken for nests.

Velvety tree ants are aggressive and when threatened will bite and spray noxious defensive secretions that resemble those of odorous house ants but are much stronger and have an intense butyric acid-like odor (Wheeler 1905). Homeowners often complain about the odor associated with an infestation of velvety tree ants.

Colonies of *L. luctuosum* are founded by multiple queens (2–40); *L. apic-*

**Figure 3.9.** Velvety tree ants (*Liometopum occidentale*) trailing on the side of a house

*ulatum* colonies have a single queen (Conconi et al. 1987). Mature colonies are large and may consist of multiple nests. Foraging trails sometimes extend 61 m (200 ft) or more before entering a house (Ebeling 1975). Files of foraging ants can be 2 to 3 cm (0.8–1.2 in) wide (Fig. 3.9) and are often hidden from view. Velvety tree ants tend homopterans for honeydew, prey on small arthropods, and scavenge on dead animals.

CONTROL

A thorough inspection is the cornerstone of a pest management program for velvety tree ants, and it should include finding nests and foraging trails (Hansen and Klotz 1999). Any dead wood such as stumps, trees, or landscape timbers should be inspected for nests. Although these ants are attracted to water-damaged wood, nests have been found in dry, sound wood associated with foam insulation. Because the ants are cryptic and more active at night, inspections after dark can be helpful.

Management can be achieved using a combination of baits and sprays. Apply baits in or near foraging trails first, then apply dusts and sprays, allowing time for the ants to carry the bait active ingredient into the nest.

"Resting" areas or satellite nests within structures may be treated with dust formulations. A perimeter spray will prevent reentry into the structure; this is particularly important if the outside nest cannot be found. Foraging trails in the surrounding landscape and on utility lines should also be sprayed, as well as the trunks of any trees where ants are foraging or may be nesting.

Tree branches in contact with the structure should be trimmed back because they may provide access for ants. Branches touching wires and cables that lead to the structure should also be trimmed because ants can gain access along these avenues as well.

## Key to Species of *Liometopum*

1 Scape of major worker surpassing the occipital corner by an amount at least twice as great as the maximum thickness of the scape; erect hairs on gastric dorsum of different lengths (Fig. 3.10a) . . . . . . . . . . *L. apiculatum*

Scape of major worker surpassing the occipital corner by an amount that does not exceed the maximum thickness of the scape; erect hairs on gastric dorsum short and of equal length (Fig. 3.10b) . . . . . . . . . . . . . . . . 2

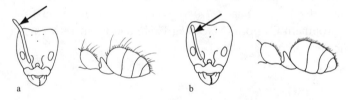

**Figure 3.10.** Head, profile of abdomen. (a) *Liometopum apiculatum* (arrow: longer scape) (b) *L. occidentale* (arrow: shorter scape)

2(1) Thick pubescence on black gaster (Fig. 3.10b); mesosoma red to yellowish . . . . . . . . . . . . . . . . . . . . . . . . . . . . . . . . . . . . . . . . . . *L. occidentale*

Pubescence on gaster sparse (Fig. 3.11); color uniformly brown . . . . . . .
. . . . . . . . . . . . . . . . . . . . . . . . . . . . . . . . . . . . . . . . . . . . . . . *L. luctuosum*

**Figure 3.11.** Abdomen profile of *L. luctuosum*

## Pyramid Ants

*Dorymyrmex* species

### IDENTIFYING CHARACTERISTICS

Workers are monomorphic with a one-segmented petiole and a sharply coni-cal protuberance on the posterior dorsal surface of the propodeum (Coovert 2005). When crushed they emit an odor of butyric acid or rotten coconut. *Do-rymyrmex insanus* is 1.5 to 2 mm long and uniformly brown; *D. bicolor* is slightly larger with an orange-brown or reddish brown head and mesosoma and a brownish black gaster (Ward 2005). These ants are called pyramid ants be-cause of their distinctive nest mounds (Coovert 2005). *Dorymyrmex insanus* is probably a complex of species (Snelling and George 1979). It is often mis-takenly called *D. pyramicus,* but that is the name of a South American species.

### DISTRIBUTION

*Dorymyrmex* species are ground-nesting ants found in open habitats at medium to low elevations (Ward 2005). *Dorymyrmex insanus* ranges from South Car-olina to North Dakota to eastern Oregon, and from its northern limit southward throughout the United States. *Dorymyrmex bicolor* is found from western Texas to southern California, Nevada, Utah, and northern Mexico (Wheeler and Wheeler 1986).

### BIOLOGY AND HABITS

Pyramid ants are also known as lion ants because of their aggressive nature (Thompson 1990); they bite but cannot sting. Conspicuous files of these fast-moving ants occasionally invade homes from outdoor nests to forage on sweets (Smith 1965). They collect nectar and honeydew and also prey on other ants and a variety of arthropods (Snelling and George 1979).

   *Dorymyrmex bicolor* colonies are polygynous and polydomous; *D. insanus* colonies vary geographically: they are monodomous and monogynous in Cal-ifornia, and polydomous and polygynous in the southeastern United States (Martinez 1995). Nests containing both species have been reported in south-ern California (Martinez 1995). Colonies, which usually consist of a few thou-sand ants, nest in dry, open, sunny areas. A crater-shaped tumulus 5 to 10 cm (1.97–3.93 in) in diameter surrounds the single entrance (Snelling and George 1979). Alates have been collected at lights in April and May (Snelling and George 1979).

CONTROL

Pyramid ants do not nest in structures but can become minor household pests when they enter homes or gardens, or forage on patios or porches (M. Martinez, pers. comm., 2005). They can be eliminated by injecting a contact insecticide into their nest entrance. Sweet baits are attractive to pyramid ants and can be used as an alternative to sprays.

**Key to Species of *Dorymyrmex***

Head and thorax deep reddish yellow, entire gaster brownish black . . . . . . . . . . . . . . . . . . . . . . . . . . . . . . . . . . . . . . . . . . . . . . . . . . . . . . . . . . . . . . . . . . . . . . . . . . . . . . . *D. bicolor*

Color variable but not as above . . . . . . . . . . . . . . . . . . . . . . . . . . . *D. insanus*

## *Forelius pruinosus*

IDENTIFYING CHARACTERISTICS

Workers are monomorphic, ranging in length from 1.8 to 2.5 mm. The antenna is twelve-segmented and without a club. The body color is highly variable, ranging from dark brown or black in the eastern United States to light brown in the western states (Ebeling 1975). When crushed they emit a rotten coconut-like odor similar to that produced by odorous house ants, which they resemble. They are also mistaken for Argentine ants, but the presence of erect hairs on the dorsal surface of the pronotum distinguishes *Forelius* from those two species (Fisher and Cover 2007).

DISTRIBUTION

This species is widely distributed from New Jersey to Florida, west to Idaho and Oregon, and south to central Mexico (Snelling and George 1979). It is common in the southern United States, where it is a household pest particularly in the Gulf Coast region (Smith 1965).

BIOLOGY AND HABITS

*F. pruinosus* is a polygynous, polydomous ant (Hölldobler 1982) that prefers to nest in fields, meadows, and bare areas, although it can also be found in open areas of woods (Smith 1965). The nests are located in soil and under rocks or bark in logs or stumps. In exposed soil, the ants often construct craters.

Mating flights occur from May to July in Florida (Smith 1965) and have been observed in April and August in California (Snelling and George 1979). Inseminated females may return to their parent colonies and then depart with brood and workers to establish new colonies (Hölldobler 1982).

Foraging workers form conspicuous trails and can tolerate high temperatures (Snelling and George 1979). They feed on live and dead insects and tend homopterans for honeydew. These ants most frequently invade houses from outdoor nests but sometimes nest indoors. They forage on most foods but prefer sweets (Smith 1965).

CONTROL

Nests should be located by following the trailing ants. The most effective treatment is to drench the nest with a residual insecticide (Hedges 1998).

# CHAPTER FOUR

# Myrmicinae

## Subfamily Characteristics

Members of this subfamily are characterized by a two-segmented petiole. The stinger is present and usually well developed, but in some species is reduced and nonfunctional as a weapon (Bolton 1994). Myrmecinae is a widespread subfamily, especially in tropical and subtropical parts of the world, and its members exhibit wide variation in morphology and habits.

### SCIENTIFIC AND COMMON NAMES

*Aphaenogaster fulva* Roger, 1863
    *A. lamellidens* Mayr, 1886
    *A. occidentalis* (Emery, 1895)
    *A. picea* (W.M. Wheeler, 1908)
    *A. subterranea valida* W.M. Wheeler, 1915
    *A. tennesseensis* (Mayr, 1862)
*Atta texana* (Buckley, 1860), *Acromyrmex versicolor* (Pergande, 1894): Leaf-
    cutting ants
*Crematogaster lineolata* (Say, 1836): Acrobat ants
    *C. ashmeadi* Mayr, 1886
    *C. laeviuscula* Mayr, 1870
    *C. californica* W.M. Wheeler, 1919
    *C. cerasi* (Fitch, 1854)
    *C. scutellaris* (Olivier, 1792)

*Messor structor* (Latreille, 1798)
    *M. andrei* (Mayr, 1886)
    *M. pergandei* (Mayr, 1886)
*Monomorium pharaonis* (Linnaeus, 1758): Pharaoh ant
    *M. destructor* (Jerdon, 1851): "Singapore ant"
    *M. floricola* (Jerdon, 1851): "Bicolored trailing ant"
    *M. minimum* (Buckley, 1867): Little black ant
*Myrmica rubra* (Linnaeus, 1758): "European fire ant"
    *M. incompleta* Provancher, 1881
*Pheidole megacephala* (Fabricius, 1793): Bigheaded ant
    *P. bicarinata* Mayr, 1870
    *P. californica* Mayr, 1870
    *P. dentata* Mayr, 1886
    *P. floridana* Emery, 1895
    *P. hyatti* Emery, 1895
    *P. obscurithorax* Naves, 1985
    *P. pallidula* (Nylander, 1849)
    *P. vistana* Forel, 1914
*Pogonomyrmex badius* (Latreille, 1802): Florida harvester ant
    *P. barbatus* (F. Smith, 1858): Red harvester ant
    *P. californicus* (Buckley, 1867): California harvester ant
    *P. maricopa* W.M. Wheeler, 1914: Maricopa harvester ant
    *P. occidentalis* (Cresson, 1865): Western harvester ant
    *P. rugosus* Emery, 1895: Rough harvester ant
    *P. salinas* Olsen, 1934
*Solenopsis invicta* Buren, 1972: Red imported fire ant
    *S. aurea* W.M. Wheeler, 1906: "Desert fire ant"
    *S. fugax* (Latreille, 1798): "European thief ant"
    *S. geminata* (Fabricius, 1804): Tropical fire ant
    *S. richteri* Forel, 1909: Black imported fire ant
    *S. xyloni* McCook, 1880: Southern fire ant
    *S. molesta* (Say, 1836): Thief ant
*Tetramorium caespitum* (Linnaeus, 1758): Pavement ant
    *T. bicarinatum* (Nylander, 1846): Guinea ant
    *T. caldarium* (Roger, 1857)
    *T. impurum* (Foerster, 1850)
    *T. insolens* (F. Smith, 1861)
    *T. simillimum* (F. Smith, 1851)
    *T. tsushimae* Emery, 1925
*Wasmannia auropunctata* (Roger, 1863): Little fire ant

## Key to Genera of Myrmicinae

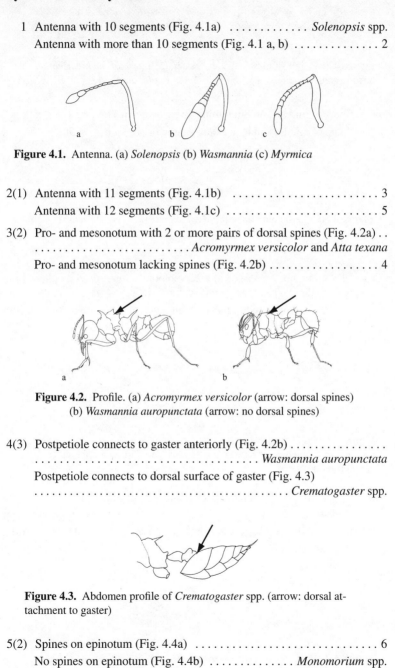

1  Antenna with 10 segments (Fig. 4.1a)  . . . . . . . . . . . . *Solenopsis* spp.
   Antenna with more than 10 segments (Fig. 4.1 a, b) . . . . . . . . . . . . . . 2

**Figure 4.1.** Antenna. (a) *Solenopsis* (b) *Wasmannia* (c) *Myrmica*

2(1)  Antenna with 11 segments (Fig. 4.1b)  . . . . . . . . . . . . . . . . . . . . . . . 3
      Antenna with 12 segments (Fig. 4.1c) . . . . . . . . . . . . . . . . . . . . . . . 5

3(2)  Pro- and mesonotum with 2 or more pairs of dorsal spines (Fig. 4.2a) . .
      . . . . . . . . . . . . . . . . . . . . . . . . *Acromyrmex versicolor* and *Atta texana*
      Pro- and mesonotum lacking spines (Fig. 4.2b) . . . . . . . . . . . . . . . . . 4

**Figure 4.2.** Profile. (a) *Acromyrmex versicolor* (arrow: dorsal spines)
(b) *Wasmannia auropunctata* (arrow: no dorsal spines)

4(3)  Postpetiole connects to gaster anteriorly (Fig. 4.2b) . . . . . . . . . . . . . . .
      . . . . . . . . . . . . . . . . . . . . . . . . . . . . . . . . . *Wasmannia auropunctata*
      Postpetiole connects to dorsal surface of gaster (Fig. 4.3)
      . . . . . . . . . . . . . . . . . . . . . . . . . . . . . . . . . . . . . *Crematogaster* spp.

**Figure 4.3.** Abdomen profile of *Crematogaster* spp. (arrow: dorsal attachment to gaster)

5(2)  Spines on epinotum (Fig. 4.4a) . . . . . . . . . . . . . . . . . . . . . . . . . . . . . 6
      No spines on epinotum (Fig. 4.4b) . . . . . . . . . . . . . . *Monomorium* spp.

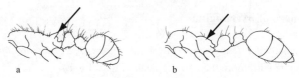

**Figure 4.4.** Profile of mesosoma and gaster. (a) *Tetramorium bicarinatum* (arrow: epinotal spines) (b) *Monomorium floricola* (arrow: no epinotal spines)

6(5)  Spurs of middle and hind tibiae finely pectinate (Fig. 4.5a)  . . . . . . . . 7
      Spurs of middle and hind tibiae simple or absent (Fig. 4.5b)  . . . . . . . 8

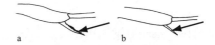

**Figure 4.5.** Spur on hind tibia. (a) *Myrmica* (arrow: pectinate spur) (b) *Tetramorium* (arrow: simple spur)

7(6)  Mesosoma not impressed at meso-epinotal suture; psammophore (long hairs under the head) present (Fig. 4.6a) . . . . . . . . . *Pogonomyrmex* spp.
      Mesosoma impressed at meso-epinnotal suture, no psammophore (Fig. 4.6b)  . . . . . . . . . . . . . . . . . . . . . . . . . . . . . . . . . . . . . . *Myrmica* spp.

**Figure 4.6.** Profile of head and mesosoma. (a) *Pogonomyrmex* spp. (arrow: no impression at epinotal suture) (b) *Myrmica* spp. (arrow: impression at epinotal suture)

8(6)  In lateral view, clypeus below antennal fossa elevated to form a ridge (Fig. 4.7a)  . . . . . . . . . . . . . . . . . . . . . . . . . . . . . . . . *Tetramorium* spp.
      In lateral view, clypeus below antennal fossa not elevated to form a ridge (Fig. 4.7b) . . . . . . . . . . . . . . . . . . . . . . . . . . . . . . . . . . . . . . . . . 9

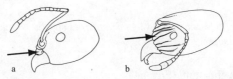

**Figure 4.7.** Profile of head. (a) *Tetramorium* spp. (arrow: elevated clypeus) (b) *Pheidole* spp. (arrow: without elevated clypeus)

9(8) Workers monomorphic, or if polymorphic, head of major not dispropor-
tionately large . . . . . . . . . . . . . . . . . . . . . . . . . . . . . . . . . . . . . . . . . . . . 10

Workers dimorphic, and head of major disproportionately large; antenna
with a distinct three-segmented club . . . . . . . . . . . . . . . . *Pheidole* spp.

10(9) Slight constriction behind postpetiole; node low and not sharply set off
from thick postpetiole (Fig. 4.8a) . . . . . . . . . . . . . . . . . . *Messor* spp.

Strong constriction behind postpetiole; node distinct and sharply set off
from postpetiole; psammophore not present (Fig. 4.8b) . . . . . . . . . . . .
. . . . . . . . . . . . . . . . . . . . . . . . . . . . . . . . . . . . . . . *Aphaenogaster* spp.

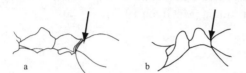

**Figure 4.8.** Petiole and postpetiole. (a) *Messor* spp. (arrow: slight constriction) (b) *Aphaenogaster* spp. (arrow: strong constriction)

## Pharaoh Ant

*Monomorium pharaonis* (Plate 3b)

### IDENTIFYING CHARACTERISTICS

Workers are about 2 mm long, monomorphic, and have a two-segmented petiole. The antenna is twelve-segmented with a three-segmented club. The body
and head are yellowish or light brown to reddish, with the tip of the gaster
darker (Vail et al. 1994). They possess a nonfunctional stinger that is used to
dispense pheromones (Edwards 1986). Queens are similar in color to workers
but about twice their size. Males are black and about 3 mm long. Pharaoh ants
are often confused with thief ants, but thief ants have a ten-segmented antenna
with a two-segmented club.

### DISTRIBUTION

Linnaeus described a specimen from Egypt in 1758. Various tropical points of
origin have been postulated for pharaoh ants, including South America, Africa,
and India (Vail and Williams 1994). They have spread throughout the world
via international trade and are probably present in every commercially important city in the United States (Smith 1965). Pharaoh ants have been established
in Europe for more than 100 years and are now found in all European countries, but in colder regions can survive only in continuously heated buildings.

Their populations have been on the increase over the last 15 years in Switzerland, particularly in apartment buildings, although their incidence in hospitals and food production facilities has been relatively low (Umwelt-und Gesundheitsschutz Zürich 2004). Reports of pharaoh ant infestations have declined markedly in Denmark over the last 15 years as a result of Internet access to management techniques and the development of more effective baits. Pharaoh ants are widely distributed in the Czech Republic and Poland, where they commonly infest apartment buildings (Czechowski et al. 2002). They have also been collected in Japan, Central and South America, Australia, New Zealand, India, and Saudi Arabia (McGlynn 1999).

## BIOLOGY AND HABITS

Pharaoh ants are a tramp species characterized by transient nesting behavior and the ability to propagate by fission or budding. Indeed, their notoriety in pest control derives from this predisposition to form satellite colonies throughout structures. Application of repellent residual insecticides can trigger the splintering of a colony into numerous satellite colonies (Vail 1997), and when such treatments are applied repeatedly, an infestation can spread throughout an entire building complex.

Even in the absence of repellent insecticides, a localized infestation of pharaoh ants can spread quickly through inadvertent transport by humans (Edwards 1986). As few as five workers and fifty pieces of brood (eggs, larvae, and pupae) can give rise to a colony of ten thousand workers in a little more than a year (Vail 1997), and these queenless propagules can produce reproductives. Mature colonies are polygynous, polydomous, and unicolonial, and mating takes place within the nest (Fig. 4.9).

The queens obtain nourishment primarily from larval secretions, and their fecundity is directly related to the number of large larvae in the nest (Fig. 4.10; Borgesen 2000). Ovariole uptake of vitellogenin is inhibited and oogenesis ceases if the queens do not obtain the larval secretions (Borgesen 2000). A group of replete workers stores the larval secretions in order to provide food for the queens when larval numbers are low (Borgesen 2000).

Pharaoh ants nest indoors in cooler climates, but in warmer climates they can forage and nest outdoors. In central Europe, for example, they are found only indoors and are spread to other buildings in items such as furniture, laundry, and electronic equipment. They are opportunistic nesters and will occupy any crack or crevice with sufficient warmth and humidity. In homes, they are often found in kitchens and bathrooms near sources of water. Typically, the nests are located in inaccessible areas, such as within interior or exterior wall

**Figure 4.9.** Multiqueen colony of pharaoh ants (*Monomorium pharaonis*) with queens, workers, and brood

voids, under floors, behind electric outlet plates in bathrooms and kitchens, and behind baseboards or windowsills. In health care facilities, they are considered a risk because they transmit disease organisms such as *Pseudomonas, Staphylococcus, Salmonella, Clostridium,* and *Streptococcus* (Beatson 1972). They may even infest intravenous units and enter sterile packages and wound dressings (Beatson 1973; Vail 1997).

Pharaoh ants forage both day and night, and often go undetected because of

**Figure 4.10.** Pharaoh ant (*Monomorium pharaonis*) worker transporting a larva

their small size and cryptic habits. Within structures, their trails are often hidden behind baseboards, cabinets, and walls, where the ants travel on wires and pipes, sometimes becoming evident when they emerge from electrical outlets and plumbing fixtures. Outside trails are often found around windows and doorways and along structural edges. Their foraging range in and around structures can be extensive. Pharaoh ants fed dyed bait in a military housing complex were tracked as far as 45 m from the bait, both inside and outside the complex, and on different floors (Vail and Williams 1994).

The principal component of their trail pheromone is faranal produced by the Dufour's gland (Hölldobler and Wilson 1990). Surprisingly, the trail is polarized and based on its geometry provides pharaoh ants with directional information (Jackson et al. 2004). This is so far the only case of trail polarity demonstrated in ants. In addition, foragers deposit a repellent pheromone on unrewarding trails to discourage other foragers from using them (Robinson et al. 2005).

Pharaoh ants are omnivorous. They feed on various sources of fat, protein, and carbohydrate (Fig. 4.11; Edwards 1986) and also scavenge or kill small insects. Pharaoh ants studied in laboratory experiments showed "alternation" and "satiation" responses to food (Edwards and Abraham 1990); that is, colonies that had been fed protein for several weeks chose carbohydrates when given a choice between the two, and those fed carbohydrates chose protein. Ants that were provided with a preferred food for several weeks switched to other foods when offered a choice. These feeding behaviors promote a balanced and varied diet for pharaoh ant colonies.

## CONTROL

Research on methods to control pharaoh ants has yielded several critical insights. Among the more significant findings is that pharaoh ant colonies fracture into satellite nests when exposed to repellent insecticides; consequently, application of repellent sprays can be counterproductive (Vail 1997). An exception to this rule is when the nests are accessible to treatment; for example, in a potted plant (D. H. Oi, pers. comm., 2004). Nonrepellent residual insecticides (e.g., fipronil and chlorfenapyr) can be used effectively in pharaoh ant control without fragmenting colonies (Buczkowski et al. 2005).

The development of baits for pharaoh ant control is a major success story. The research laboratories of J. P. Edwards with the Ministry of Agriculture in England and D. F. Williams with the USDA-ARS in Gainesville, Florida, have made substantial contributions to our understanding of pharaoh ant behavior and its importance in bait efficacy. Oi et al. (2000), for example, showed that

**Figure 4.11.** Pharaoh ant (*Monomorium pharaonis*) workers at a food source

bait containing an insect growth regulator (pyriproxyfen) is distributed by the ants more thoroughly between colonies than bait containing a metabolic inhibitor (hydramethylnon). Thus, more bait stations are necessary when metabolic inhibitors are used. A thorough and systematic application of baits is required for effective control. In structural infestations in which pharaoh ants are foraging both inside and outside, outdoor applications of bait are sufficient (Oi et al. 1994). Baiting programs should include procedures to monitor changes in food preference (Edwards and Abraham 1990).

The first step in a comprehensive baiting program should be a pretreatment survey to determine where the ants are located. White index cards dabbed with peanut butter or honey and placed at potential food and water sources are easy to use and effective. If you're investigating multiple locations, an infested apartment complex, for instance, a 60 cc syringe with the needle tip removed makes applications more convenient and less messy (Vail 1997). In cooler climates the baited cards should be placed inside the structure; in warmer climates they should be placed both inside and outside. Thorough coverage is essential. In an apartment complex, for example, survey cards should be placed in apartment units and in lobbies, kitchens, laundries, lounges, and offices. Each unit should have at least eight to sixteen cards (usually two to four each in the kitchen, living room, bathroom, and bedroom). Survey cards should also be placed where tenants have seen ant activity. Windowsills in the living room and bedroom of each apartment are ideal spots for survey cards because dead insects on windowsills attract foraging ants, and the numerous cracks and crevices around window casings are ideal nest sites. In kitchens and bathrooms, cards should be placed near sources of water such as pipes and drains, sinks, countertops, and toilets. Outside survey cards should be placed near windows, around entrances, and near water lines. It is important to position the cards along edges or other structural guidelines where the ants are likely to travel. Allow one to several hours for ants to locate and recruit foragers to the attractant and then estimate the number of worker ants on each card and record the numbers on an inspection diagram.

The pretreatment survey locates where the ant activity is concentrated so that toxic baits can be placed strategically. It will also help to find small, isolated colonies that otherwise might be overlooked and could result in continued infestation. At sites where feeding is recorded, the trailing ants can be traced back to their entrance site—a crack or crevice, switch plate, or pipe flange, for example. Since pharaoh ants typically travel on wires and pipes, baiting every switch plate and pipe void in each room with ant activity is recommended. Baiting outside is important because a considerable number of pharaoh ants forage outdoors during warm weather.

A good floor plan of the infested structure is useful to display survey data and locate bait placements. A floor plan also helps in estimating the quantity of bait or number of bait stations needed. Four to six bait placements may be sufficient for a room of average size, but it is important that every ant trail be baited, so additional placements may be necessary. A year or longer may be required to control an infestation in a building with a long history of pharaoh ant activity combined with attempts at control using repellent residual insecticides.

Success is likely when ants are attracted to the bait and are induced to feed on it for an extended period, but pharaoh ants have a tendency to switch their food preferences, so it is important to provide choices in the form of sweet liquid baits and a variety of protein or oil baits. Foragers may tire of one kind of food and switch to another if offered a choice.

### Key to Species of *Monomorium*

    1 Dimorphic workers; epinotum transversely rugulose (finely wrinkled) (Fig. 4.12a) . . . . . . . . . . . . . . . . . . . . . . . . . . . . . . . . . . . . . *M. destructor*
    Monomorphic workers; epinotum not rugulose (Fig. 4.12b) . . . . . . . . 2

**Figure 4.12.** Dorsal view of petiole, postpetiole, and epinotum. (a) *M. destructor* (arrow: rugae on epinotum) (b) *M. pharaonis* (arrow: no rugae on epinotum).

2(1) Head and gaster dark brown, thorax lighter . . . . . . . . . . . . . *M. floricola*
Color not as above . . . . . . . . . . . . . . . . . . . . . . . . . . . . . . . . . . . . . 3

3(2) Body dark brown or black, and smooth and shiny . . . . . . . . *M. minimum*
Body light brown or yellowish; head, mesosoma, and petiole punctate
. . . . . . . . . . . . . . . . . . . . . . . . . . . . . . . . . . . . . . . . . . . . . *M. pharaonis*

## Little Black Ant and Related Species

*Monomorium* Species (Plate 3c)

### IDENTIFYING CHARACTERISTICS

These are small ants that have a twelve-segmented antenna with a three-segmented club. Workers of *M. minimum* and *M. floricola* are monomorphic and 1.5 to 2 mm and 1.4 to 1.8 mm long, respectively (Smith 1965). Workers of *M. destructor* are somewhat dimorphic and 1.8 to 3 mm long (Smith 1965). As the common name implies, *M. minimum* workers are black, and they have a shiny appearance (Snelling and George 1979). Queens are similar in color but larger, about 4 mm long (Thompson 1990). The "bicolored trailing ant" (*M. floricola*) resembles the little black ant but is more slender and has a dark brown or blackish head and gaster and a lighter mesosoma, petiole, and legs (Smith 1965). The "Singapore ant" (*M. destructor*) is pale yellowish or very light brown with variable amounts of black on the gaster (Vail et al. 1994).

### DISTRIBUTION

The little black ant is native to North America and is found in southeastern Canada, the northeastern United States to the Pacific Coast, and south into Mexico (Snelling and George 1979; Wheeler and Wheeler 1986). It is a common structural pest in the south-central United States from western Tennessee to central Texas (Hedges 1997) but is rare in the Pacific Northwest (Snelling and George 1979). The other two species found in North America, *M. floricola* and *M. destructor,* are tramp species that originated in India and Southeast Asia, and Africa or India, respectively (McGlynn 1999). *Monomorium floricola* is widespread in Florida, where it is a minor structural pest (Vail et al. 1994), and in Hawaii, where it is among the most commonly encountered ants in buildings (Reimer et al. 1990). *Monomorium destructor* is common in Key West (Deyrup 1991). It is also found in several other counties in Florida, where it is a minor structural pest (Vail et al. 1994), and in Tennessee (Smith 1965). In Europe, *M. floricola* was intercepted in Switzerland three times in 2004 and 2005, and once in Hamburg, Germany, in a shipment of plants from Haiti (Sellenschlo 2002a).

## Biology and Habits

Colonies of all three species of *Monomorium* are moderate to large and polygynous. A nest of *M. minimum* in Nevada, for example, had thirty-one queens (Wheeler and Wheeler 1986). Laboratory studies showed that the growth potential of colonies of *M. floricola* as measured by intrinsic rate of increase is similar to that of pharaoh ants, and that both species are capable of producing new queens if at least one thousand eggs and one hundred workers are present (Eow et al. 2004). Growth of *M. destructor* colonies is significantly slower.

*Monomorium floricola* appears to multiply by budding alone, while *M. minimum* and *M. destructor* may also engage in mating flights. Alates of *M. minimum* have been observed from June to August (Smith 1965).

Little black ants nest in soil in the open or under objects, and in decaying wood (Snelling and George 1979). Nests of *M. floricola* are primarily arboreal in twigs, branches, or under bark; *M. destructor* nests are usually in soil or inside buildings (Smith 1965).

All three species tend homopterans for honeydew and feed on insects (Smith 1965). Little black ants also collect plant exudates and pollen, and may kill hatchling birds (Smith 1965). Indoors, *M. minimum* and *M. destructor* feed on a variety of foods including sweets, meats, grease, bread, oils, cornmeal, fruits, and fruit juices; little is known about the food preferences of *M. floricola* (Smith 1965). In laboratory tests with starved ants, *M. floricola* workers preferred a lipid-based food (peanut oil) to carbohydrate (10% sucrose water) and protein (canned tuna), while *M. destructor* workers preferred the carbohydrate and protein-based foods (Eow et al. 2005).

*Monomorium minimum* workers use a pheromone to recruit their nestmates to food sources (Adams and Traniello 1981), with the number of ants following a trail being a linear function of the amount of pheromone deposited (Wilson 1971). The trail pheromone is produced in the Dufour's gland (Adams and

Traniello 1981). The ants defend clumped resources by "gaster-flagging," a behavior in which the gaster is vibrated to disperse venom on the extruded stinger (Adams and Traniello 1981). They tolerate high temperatures (Adams and Traniello 1981) and thus can forage at times when other ants are less active.

Once among the most common household pest ants in the United States (Smith 1965), little black ants have been largely displaced by invasive species such as red imported fire ants and Argentine ants (Alder and Silverman 2005; Keck et al. 2005). The other two species of *Monomorium* are important urban pests in the tropics but only minor structural pests in the United States and Europe.

CONTROL

The most effective treatment for all of the pest *Monomorium* species is to locate the nest and treat it with a residual insecticide (Hedges 1998): a water-based formulation for outdoor nests, and dusts or aerosols for nests in structural voids. Most household infestations of little black ants originate outdoors and can be traced to a stump, tree, log, fence, or pile of lumber or bricks; they also nest in landscape mulch and under stones (Hedges 1997). Indoors, colonies may nest in the voids of walls and cabinets, as well as within and behind foundations and brick or stone veneer (Hedges 1997).

If the nest cannot be found, baits are an effective alternative, although the choice of bait and its application should be carefully considered with respect to the colony's food preferences and whether other species of ants are present. For example, in laboratory tests, an *M. minimum* colony decreased its consumption of toxic bait when forced to compete with Argentine ants (Alder and Silverman 2005). Changing the bait base and timing the bait placements may minimize this competitive interference in the field. *Monomorium minimum* workers preferred a protein-based bait and were active only during the day, for example, while *L. humile* workers preferred sugar-based bait and were active both day and night.

## Bigheaded Ant and Related Species

*Pheidole* Species (Plate 3d)

IDENTIFYING CHARACTERISTICS

Workers have a two-segmented petiole, a twelve-segmented antenna with a three-segmented club, and are of two types: slender minors and more robust, large-headed majors (Wilson 2003). Minors range in size from 2 to 2.8 mm, and majors are 3.5 to 4.5 mm. Workers have propodeal spines and a stinger. Workers of *Pheidole megacephala* are brownish yellow, and the majors have large, heart-shaped heads (Wilson 2003). Bigheaded ants are easy to identify when majors can be collected (peanut butter can be used to attract them), but minors are more numerous in the colony and resemble fire ants (Vail et al. 1994). Male and female reproductives are 4.6 to 6 mm long.

## DISTRIBUTION

The genus *Pheidole* is possibly the largest and most diverse of the ant genera, with 624 described species in the New World alone (Wilson 2003). The only other genus close to its size is *Camponotus,* with about 400 species in the New World (Wilson 2003). The major pest in this group is the bigheaded ant (*P. megacephala*), also known as the "coastal brown ant," an invasive tramp species of African origin now widely distributed in the New World, Europe, · Australia, the Arabian Peninsula, and various Atlantic and Pacific islands (McGlynn 1999; Wilson 2003). Its introduction into Europe was first described in Spain. *Pheidole pallidula* is established in the Mediterranean region and is found as far north as Switzerland. It has been intercepted several times in Berlin inside luggage from Asia Minor and has also been accidentally introduced into Florida and Canada. A colony of an unidentified species of *Pheidole* was intercepted in Wiesbaden, Germany, in a shipment of household equipment from the United States (Sellenschlo 2002a). Unidentified species of *Pheidole* have been twice intercepted in Switzerland (Umwelt- und Gesundheitsschutz Zürich 2004), and a large, established *Pheidole* colony was found in a centrally heated apartment bloc near London, England (C. Boase, pers. comm., 2006).

PMPs collected seven species of *Pheidole* in and around homes in a survey in Florida (Klotz et al. 1995), with *P. megacephala* the most common. In Hawaii, *P. megacephala* is a major pest in pineapple groves and occasionally in structures (G. Taniguchi, pers. comm., 2005). *Pheidole bicarinata, P. floridana,* and *P. dentata* infest houses in the eastern United States (Smith 1965). *Pheidole hyatti* and *P. vistana* have been reported as household pests in the western United States, and *P. californica* was collected in a greenhouse (Ebeling 1975; Wheeler and Wheeler 1986). *Pheidole obscurithorax* is a South American species that was introduced into the United States around 1950. It has been collected in Alabama and Florida and is continuing to increase its range (Storz and Tschinkel 2004). It coexists with *S. invicta* and can tolerate highly disturbed habitats.

**Figure 4.13.** Bigheaded ant (*Pheidole megacephala*) infestation in the bathroom of a building

## BIOLOGY AND HABITS

Most of the North American species form small colonies with two hundred to three hundred individuals and nest in soil (Creighton 1950), but colonies of *P. megacephala* can reach tremendous size, with huge, extended nests and multiple queens (Hedges 1997). They thrive in moist, disturbed habitats in both urban and agricultural settings (Wilson 2003). These supercolonies are highly competitive and territorial, and displace other species of ants.

Brood development requires temperatures above 25 °C combined with high humidity, and establishment of outdoor colonies in central Europe thus seems unlikely. Infestations in Germany, for example, are only found in buildings that maintain a constant high temperature. Indoors, these ants are commonly found in bathrooms and kitchens near sources of water (Fig. 4.13).

Most species feed on seeds, which the majors crack with their massive jaw muscles (Creighton 1950). They feed voraciously on animal tissue when the opportunity arises (Creighton 1950), however, and certain species are predators of fire ant queens (Bhatkar 1990). Honeydew is less attractive to many species (Creighton 1950), although *P. megacephala* is a serious pest in pineapple groves in Hawaii because of its association with mealybugs.

Workers forage for proteins, taking meat, fish, cheese, peanut butter, and dog and cat food, but also fruit juice. They forage in garbage cans and on animal

**Figure 4.14.** Bigheaded ant (*Pheidole megacephala*) workers feeding on a dead American cockroach

carcasses and other decaying organic matter, so there is always a risk of disease transmission if they are present (Fig. 4.14). Some species are intermediate hosts for tapeworms of wild and domesticated turkeys and other fowl (Smith 1965).

## CONTROL

Bigheaded ants often nest next to foundations, where they may construct mud tubes that can be mistaken for subterranean termite galleries (Hedges 1997). Outdoors, their foraging trails are easily located and traced back to the nest, which can be treated directly with insecticidal dusts or water-based sprays. Foraging workers sometimes travel up to 50 m between their nest and a food source in a building, and invasion of upper floors is possible via branches of trees and shrubs. Indoors, protein-rich food sources, cages for pets, and potted plants should be inspected.

In peninsular Florida and Key Largo, extensive infestations of *P. megacephala* can be found around buildings, ornamental plant bases, sidewalks, and driveways (Deyrup 1991; Klotz et al. 1995); one supercolony covered an entire block (Mangold 1996). Infestations typically originate outside a structure or beneath a slab foundation. Potted plants are also common nest sites, and when brought indoors can transport colonies inside. Owners of infested buildings typically complain about finding hundreds of ants, both live and dead. Control is difficult because there is a high probability of reinfestation if the entire supercolony is not treated.

Protein-based baits with peanut butter or liver attract bigheaded ants. *Pheidole megacephala* infestations in pineapple fields were controlled with Amdro® (1% hydramethylnon) delivered in bait stations (Taniguchi et al. 2005). Protein-based hydramethylnon granular bait delivered in stations eliminated an infestation of *P. megacephala* in a warehouse (G. Taniguchi, pers. comm., 2005).

## Key to Species of *Pheidole*

1　Node low and weakly developed; mesonotum not convex (Fig. 4.15a) . . . 2
　Node strongly developed; mesonotum convex (Fig. 4.15b) . . . . . . . . . . 3

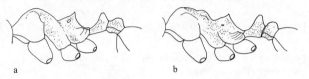

**Figure 4.15.** Mesosoma, petiole, and postpetiole. (a) *P. megacephala* (arrow: low node, mesonotum) (b) *P. californica* (arrow: developed node, convex mesonotum)

2(1)　Major workers with postpetiole as broad as long (Fig. 4.16a) . . . . . . . . .
　. . . . . . . . . . . . . . . . . . . . . . . . . . . . . . . . . . . . . . . . . . . . . . . . . *P. megacephala*
　Major workers with postpetiole wider than long (Fig. 4.16b) . . . . . . . . . .
　. . . . . . . . . . . . . . . . . . . . . . . . . . . . . . . . . . . . . . . . . . . . . . . . . . . *P. pallidula*

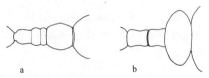

**Figure 4.16.** Dorsal view of petiole and postpetiole. (a) *P. megacephala* (arrow: postpetiole) (b) *P. pallidula* (arrow: wide postpetiole)

3(1)　Head of major worker quadrate or rectangular with sides parallel, broadest anterior to the occipital lobes (Fig. 4.17a) . . . . . . . . . . . . . . . . . . . . 4
　Head of major worker gradually narrows toward mandibular insertions, broadest at the occipital lobes (Fig. 4.17b) . . . . . . . . . . . . . . . . . . . . . . 5

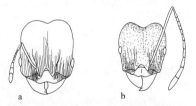

**Figure 4.17.** Face view. (a) *P. bicarinata* (b) *P. vistana*

4(3)  Occipital lobes of majors sculptured (Fig. 4.18) . . . . . . . . *P. californica*
      Occipital lobes of majors smooth (Fig. 4.17a) . . . . . . . . . . *P. bicarinata*

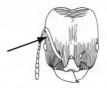

**Figure 4.18.**  Face view of *P. californica*

5(3)  Scape of major short, extending at most slightly beyond midpoint between
      eye and occipital corner (Fig. 4.19a) . . . . . . . . . . . . . . . . . . *P. floridana*
      Scape of major extending well beyond the midpoint between eye and oc-
      cipital corner (Fig. 4.19b) . . . . . . . . . . . . . . . . . . . . . . . . . . . . . . . . . . . 6

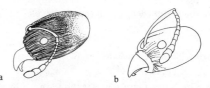

**Figure 4.19.**  Profile of head. (a) *P. floridana* (b) *P. dentata*

6(5)  In face view, occipital lobes sculptured (Fig. 4.20a) . . . . *P. obscurithorax*
      In face view, occipital lobes smooth (Fig. 4.20b) . . . . . . . . . . . . . . . . . 7

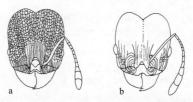

**Figure 4.20.**  Face view. (a) *P. obscurithorax* (b) *P. dentata*

7(6)  Antennal scapes of major worker not reaching the occipital corners (Fig.
      4.20b) . . . . . . . . . . . . . . . . . . . . . . . . . . . . . . . . . . . . . . . . . . . . . . *P. dentata*
      Antennal scapes of major worker reaching occipital corners (Fig. 4.21) . .
      . . . . . . . . . . . . . . . . . . . . . . . . . . . . . . . . . . . . . . . . . . . . . . . . . . . . . . . . . . 8

8(7)  Antennal scapes exceed occipital corners by five times their maximum
      width (Fig. 4.17b) . . . . . . . . . . . . . . . . . . . . . . . . . . . . . . . . . . . . *P. vistana*
      Antennal scapes exceed occipital corners by no more than half their max-
      imum width (Fig. 4.21) . . . . . . . . . . . . . . . . . . . . . . . . . . . . . . . . *P. hyatti*

**Figure 4.21.** Face view of *P. hyatti*

## Pavement Ant and Related Species

*Tetramorium* Species (Plate 3e)

### Identifying Characteristics

Workers of *Tetramorium* are monomorphic and have a twelve-segmented antenna with a three-segmented club. They range in length from 2.5 to 3 mm, bear a pair of propodeal spines of variable length, and have enlarged femora (Vail et al. 1994).

The species referred to as the pavement ant, *T. caespitum,* in the United States is small, ranging in size from 2.5 to 3 mm long. The dorsal surface of the head and thorax is sculptured with numerous parallel grooves, and the body color varies from brown to dark red-brown to black (Merickel and Clark 1994); the legs and antennae are lighter. Queens are dark brown, 6 to 8 mm long, and also have a sculptured head and thorax and a pair of propodeal spines. Males are slightly smaller (5.7–7 mm long) and lack the propodeal spines. According to Trager (pers. comm., 2007) this species should be called by the temporary label *Tetramorium* species E.

Workers of the Guinea ant (*T. bicarinatum*) are 3 to 4 mm long and resemble *Solenopsis xyloni* workers (Martinez 1993). They have a pale red to bright orange-brown head and mesosoma and a much darker brown gaster (Martinez 1993). Queens are similar in color but larger (4.5–5 mm long), and males are 3 to 4 mm long with a brown head and mesosoma and dark brown gaster (Martinez 1993). A new invasive species that probably originated in Japan, *Tetramorium tsushimae,* has been found in parts of Missouri and Illinois (Steiner et al. 2006). Morphologically it is difficult to differentiate from *Tetramorium* species E, but it is somewhat smaller and its propodeal spines are longer.

### Distribution

The pavement ants are a widespread group of species native to the Palaearctic region (Seifert 1996; Czechowski et al. 2002). Their taxonomic status and distribution are uncertain at this time and are likely to change in the future. Brown

(1957) and Smith (1965) proposed a European or Asian origin for pavement ants, while Creighton (1950) considered the group native to North America. The range includes most of the United States, and pavement ants are particularly common in the Midwest and along the West Coast (Bennett et al. 1997). In the Northeast and Midwest, they are the major ant pests in commercial buildings (Hedges 1997). Pavement ants are found in almost every metropolitan area in Washington and Oregon, and are significant urban pests in Idaho as well (Merickel and Clark 1994). They are also found on the Hawaiian Islands and in Brazil and Japan (McGlynn 1999).

In Europe, pavement ants are seldom found in buildings, although infestations in apartment complexes have been reported in one town in the Czech Republic. The species that occurs there is much more often found outdoors inhabiting plains and low mountain areas, typically in sandy soil. A closely related species, *T. impurum,* is found in loamy soil in hilly and mountainous areas (Seifert 1996).

Four widespread tropicopolitan tramp species of *Tetramorium* are sporadically found in European and North American tropical greenhouses: *T. insolens* and *T. bicarinatum,* which probably originated in the Pacific and the Orient, respectively, and *T. caldarium* and *T. simillimum,* which originated in Africa (Czechowski et al. 2002; M. Martinez, pers. comm., 2007).

The Guinea ant (*T. bicarinatum*) is a global tramp species that has been widely disseminated throughout the tropics by commerce (Martinez 1993). There are records of this ant in greenhouses from several parts of the United States (Creighton 1950), as yard pests in several counties in Florida (Vail et al. 1994), and as minor household pests in Long Beach, California (M. Martinez, pers. comm., 2005). A related exotic tramp species found in greenhouses in the United States and as yard pests in Florida is *T. simillimum* (Creighton 1950; Vail et al. 1994), which has also made its way to Madagascar, Australia, and the Arabian Peninsula (McGlynn 1999).

BIOLOGY AND HABITS

As a tramp species the pavement ant prefers disturbed habitats associated with human activity (McGlynn 1999). Colonies are monogynous (Bruder and Gupta 1972; Seifert 1996), and one of the authors (LH) has observed pavement ant wars in Washington State. Polygynous colonies have been reported in Austria in alpine meadows (Steiner et al. 2003). Colonies have a high reproductive capacity and the ability to live in a wide variety of climates and feed on diverse foods (Merickel and Clark 1994).

The pavement ant derives its name from its habit of nesting beside and under sidewalks, driveways, rocks, and foundations; a conspicuous pile of excavated soil is the telltale sign of a nest. Nests are also located in open areas or under stones, and in masonry or rotting wood (Smith 1965). During winter, these ants will move inside, preferably to be near a heat source such as a radiator or heating duct (Hedges 1997). Workers often enter buildings through expansion joints, cracks in a slab, and openings for utilities.

New colonies are established after mating flights. The flights usually take place in the spring or early summer but may occur at other times of the year if the ants are nesting indoors (Hedges 1997). The alates typically emerge from under baseboards, expansion joints, or from floor registers connected to heating ducts. In commercial buildings, they often become a nuisance when the alates emerge from openings in walls above false ceilings and then drop into the rooms below (Hedges 1992). Alates may continue to emerge for weeks.

Pavement ants tend homopterans, especially subterranean forms, for honeydew and also feed on live and dead insects and a variety of plants (Smith 1965). As household pests they are attracted to both greasy and sweet foods (Bennett et al. 1997), and are often seen foraging inside structures and around garbage cans and pet food (Merickel and Clark 1994).

Guinea ant colonies are polygynous, moderate to large, and spread by budding (Martinez 1993). Nests are usually in exposed soil or under stones, in rotting wood, in stems or branches, and under bark (Smith 1965). Guinea ants are predators and scavengers as well as seed eaters (Martinez 1993). As household pests they feed on fruits, vegetables, meats, and grease (Smith 1965) but are probably of little or no economic significance (Martinez 1993).

## CONTROL

In tests, perimeter sprays of fipronil (0.06%), imidacloprid (0.06%), and cyfluthrin (0.05%) applied 1 m (3.3 ft) out from and 0.7 m (2.3 ft) up the foundation wall were effective against pavement ants (Scharf et al. 2004). The fipronil spray was the most effective, with only 2% of the pretreatment number of ants remaining active after 8 weeks.

A number of commercially available ant baits are effective against pavement ants (Hedges 1998). These include both sugar- and protein-based baits (G. Wegner, pers. comm., 2004) and the corn grit–soybean oil baits (Oi 2002). For Guinea ants, Dash et al. (2005) recommended sweet liquid baits and possibly applying an insecticidal barrier around the structure. A drill-and-treat termiticide application using a termiticide labeled for subslab treatment of ant colonies may be warranted if pavement ants have infested the soil beneath a foundation (Hedges 1998).

## Key to Species of *Tetramorium*

1  Frontal carinae short, terminating at level of eyes (Fig. 4.22a) . . . . . . . 2
   Frontal carinae long, terminating far beyond upper level of eyes (Fig.
   4.22b) . . . . . . . . . . . . . . . . . . . . . . . . . . . . . . . . . . . . . . . . . . . . . . . . . . . . . . 3

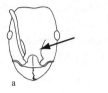

**Figure 4.22.** Face view. (a) *T. caespitum* (arrow: short frontal carinae) (b) *T.
bicarinatum* (arrow: long frontal carinae)

2(1)  Petiole and postpetiole mainly smooth and shiny, with occasional striations
      (Fig. 4.23a) . . . . . . . . . . . . . . . . . . . . . . . . . . . . . . . . . . . *T. caespitum*
      Petiole and postpetiole densely punctured, irregularly striate, rugulose
      (Fig. 4.23b) . . . . . . . . . . . . . . . . . . . . . . . . . . . . . . . . . . . . . *T. impurum*

**Figure 4.23.** Dorsal view of petiole and postpetiole. (a) *T. caespitum* (arrow:
smooth surface (b) *T. impurum* (arrow: punctate surface)

3(1)  Epinotum with long spines (Fig. 4.24a) . . . . . . . . . . . . . . . . . . . . . . . . . 4
      Epinotum with short, acute spines (Fig. 4.24b) . . . . . . . . . . . . . . . . . . 5

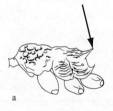 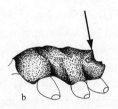

**Figure 4.24.** Mesosoma. (a) *T. bicarinatum* (arrow: long spines) (b) *T. simillimum* (arrow: short spines)

4(3)  Mandibles smooth except for hair pits; body yellow (Fig. 4.25a) . . . . . . .
      . . . . . . . . . . . . . . . . . . . . . . . . . . . . . . . . . . . . . . . . . . . . . . . . . . . *T. insolens*
      Mandibles densely and finely striate; mesosoma yellow, gaster dark (Fig.
      4.25b) . . . . . . . . . . . . . . . . . . . . . . . . . . . . . . . . . . . . . . . . . . *T. bicarinatum*

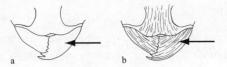

**Figure 4.25.** Face view of mandibles. (a) *T. insolens* (arrow: smooth mandibles) (b) *T. bicarinatum* (arrow: striate mandibles)

5(3) Frontal carinae strongly developed to nearly the occipital margin; antennal scrobes well developed (Fig. 4.26a) . . . . . . . . . . . . . . *T. simillimum*

Frontal carinae strongly developed to level of eye, then becoming weak or broken; antennal scrobes vestigial (Fig. 4.26b) . . . . . . . . . . *T. caldarium*

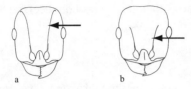

**Figure 4.26.** Face view. (a) *T. simillimum* (arrow: long frontal carinae) (b) *T. caldarium* (arrow: frontal carinae short)

## Little Fire Ant

*Wasmannia auropunctata* (Plate 3f)

### IDENTIFYING CHARACTERISTICS

Workers are monomorphic, 1.2 to 1.5 mm long, and light to golden brown. The antenna has eleven segments and a three-segmented club that appears to have only two segments because the third is so small (Smith 1965). The propodeum bears a pair of prominent spines. Queens are 4.5 to 5 mm long and dark brown (Ulloa-Chacon and Cherix 1990).

### DISTRIBUTION

From its point of origin in the Neotropics the little fire ant has been introduced into Florida, the Hawaiian Islands, Africa, and the Galápagos Islands, where it may pose a threat to wildlife (Ulloa-Chacon and Cherix 1990; McGlynn 1999). The little fire ant is occasionally reported in southern California but has not become established there, probably because the climate is unsuitable (Ward 2005).

Biology and Habits

The colony dynamics of little fire ants are similar to those of other tramp species such as Argentine ants and pharaoh ants (Ulloa-Chacon and Cherix 1990). Colonies are polygynous and spread by budding, and nests are diffuse and are interconnected by trails. The little fire ant is an opportunistic nester. Outdoor nests are typically at ground level under bricks, stones, leaves, wood, and rubbish, or around plant roots; in buildings they tend to be in structural cracks and crevices (Fernald 1947; Ulloa-Chacon and Cherix 1990). Little fire ants are considered household pests because they contaminate food, infest bedding and clothes, and inflict painful stings (Smith 1965).

Little fire ants are omnivores; they prefer honeydew but also feed on both live and dead arthropods, seeds, flowers, and leaves (Ulloa-Chacon and Cherix 1990). As household pests they feed on bacon, beef, peanut butter, oils, milk, and orange juice (Vail et al. 1994).

Control

An ongoing research program on the Galápagos Islands is aimed at eliminating *W. auropunctata,* which is not only a threat to biodiversity on the islands but also a serious structural pest on the island of Santa Cruz. Laboratory studies determined that peanut butter and soybean oil were highly attractive to little fire ant workers in food preference tests, and baits containing soybean oil and hydramethylnon (Amdro®), and peanut butter and sulfluramid (Raid Max® Ant Bait) were highly preferred in field tests (Williams and Whelan 1992). Soybean oil and hydramethylnon bait killed 100% of small laboratory colonies within 20 days.

The largest eradication program on record for little fire ants commenced in 2001 on Marchena Island in the Galápagos Archipelago (Causton et al. 2005). The infested area, about 21 ha (52 ac) in size, was treated with up to three applications of Amdro®. Only two small residual populations were detected and treated in 2002, and no little fire ants were detected in 2003 and 2004.

## Harvester Ants

*Pogonomyrmex* Species (Plate 4a)

Identifying Characteristics

Workers are 4.5 to 13 mm long (Hedges 1998), have a two-segmented petiole, and can sting. *Pogonomyrmex* means "bearded ant," a reference to the psammaphore, four fringes of long hairs located on the posterior surface of the head behind the mouthparts on some—but not all—species of harvester ants (Taber 1998). The structure is used for carrying seeds when foraging and sand or soil during nest excavation (Wheeler and Wheeler 1986). Members of all species have a pair of propodeal spines, and their color varies from red to brown to black.

## DISTRIBUTION

Twenty-three species of harvester ants are found in North America. In the United States, they commonly inhabit arid grasslands and deserts of the western states. Only six species are considered pests (Taber 1998), primarily in agriculture: the red harvester ant (*P. barbatus*) is found in Kansas, south through Texas, and into Arizona; the maricopa harvester ant (*P. maricopa*) occurs from west Texas into southern California; the western harvester ant (*P. occidentalis*) is found at high elevations (2,743 m [9000 ft]) in western states as far north as Montana and North Dakota; the California harvester ant (*P. californicus*) is among the most common ants in the Colorado and Mojave deserts (Snelling and George 1979); the "owyhee harvester ant" (*P. salinas*) is found in Utah north into Montana and the Pacific Northwest; and the Florida harvester ant (*P. badius*) is the only harvester ant found east of the Mississippi River. All are reddish brown except *P. barbatus,* which is reddish brown to red. *Pogonomyrmex rugosus,* which is not on Taber's (1998) list of pest species but has caused allergic reactions to its sting, is found from western Texas and Oklahoma westward to southern California and is generally brownish black to brownish red.

## BIOLOGY AND HABITS

Harvester ants are equally well known for their diet of seeds and their painful stings. As pests their most significant impact is in agriculture, where they cause damage to crops, rangelands, and livestock (Taber 1998). They are occasional pests in lawns and playgrounds, where children are particularly vulnerable to their stings. The stings of harvester ants are considered the most persistently painful of all North American ant stings, and the venom they contain the most toxic of all insect venoms (Schmidt 2003). During a single year, eight patients stung by the maricopa harvester ant and the rough harvester ant (*P. rugosus*) in Tucson, Arizona, sought treatment (Pinnas et al. 1977). One patient suffered several allergic reactions to stings by *P. rugosus* when an infestation in his yard was not eliminated (Pinnas et al. 1977). Allergic reactions to harvester ant stings are reported on a regular basis in Arizona (Klotz et al. 2005), and two deaths in Oklahoma were attributed to stings of the red harvester ant. Wildermuth and Davis (1931) reported red harvester ants to be "a great annoyance around dooryards and . . . especially troublesome . . . in city lawns."

The California harvester ant is among the most infamous for its sting, which—unusual for an ant—remains in the wound (Wheeler and Wheeler 1986). Herms (1939) reported that California harvester ants "will readily attack humans and smaller animals. Hog raisers in the Imperial Valley, California, report many young pigs killed by ants, particularly by the stings of *P. californicus*. It is a matter of common observation to see a small pig walk leisurely upon an ant mound and suddenly begin to kick and squeal, due to the terrific attack of the myriads of ants rushing forth from the nest. The animals commonly topple over with legs outstretched and death may result." The California harvester ant nests in exposed sandy soil, and a nest may have as many as ten entrances (Snelling and George 1979).

A colony of *P. californicus* observed in Pasadena, California, remained inside the nest for 3 or 4 months during the winter, and closed the nest entrance every night during the rest of the year. The ants showed maximum activity when ground temperatures were between 32 and 46 °C (90–115 °F) and generally remained inside the nest when temperatures were above 49 °C (120 °F). Mating flights took place in the late morning hours on clear, hot days in June and July (Michener 1942).

During such flights, which are usually triggered by a weather cue, winged male and female reproductives emerge from nests distributed over a large geographic area and fly to a congregation site, often the same location every year. A conspicuous landmark such as a bush, tree, or hilltop often marks the congregation site; *P. rugosus* mating swarms take place over a flat area in the desert (Hölldobler and Wilson 1990). After mating, the newly fertilized queens search for suitable nest sites. Each queen removes her wings and digs a chamber in the ground where she lays many eggs, most of which serve as food for her or the faster-developing larvae. She also obtains nutrition from her fat reserves and her now-nonfunctional wing muscles, which are broken down and resorbed. The first workers to emerge care for the subsequent brood, expand the nest, and begin foraging. Hereafter, the sole function of the queen is to lay eggs. After attaining considerable strength in numbers, the colony produces a new group of alates.

Harvester ant colonies are monogynous, founded by a single queen, and may survive for 15 to 20 years. Red harvester ants create, large circular nest clearings, denuding vegetation and contributing to soil erosion. Their nests may be 1.8 m (6 ft) deep and sometimes damage roads and airport runways by causing potholes and erosion (Taber 1998).

The western harvester ant constructs the most complex nest of all the harvesters; nest mounds can be 1 m (3.3 ft) tall, 4.8 m (15.7 ft) in diameter, and 6 m (19.7 ft) deep, and nests may contain as many as ten thousand ants (Taber 1998). The large mounds can interfere with mowing and harvesting equipment, and colonies can strip away as much as 20% of the vegetation in pastures (Taber 1998).

## CONTROL

Corn grit–soybean oil baits developed for red imported fire ants are effective against harvester ants (Wagner 1983). Activity in nests of *P. californicus* and *P. rugosus* respectively ceased within 48 hours and 2 to 3 weeks of treatment with hydramethylnon bait (Wagner 1983).

### Key to Species of *Pogonomyrmex*

1 Worker caste strongly polymorphic, major with disproportionately enlarged head (Fig. 4.27a) .............................. *P. badius*
    Worker caste not polymorphic; head not disproportionately enlarged (Fig. 4.27b) ................................................ 2

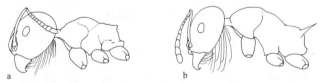

**Figure 4.27.** Head and mesosoma. (a) *P. badius* (b) *P. occidentalis*

2(1) Distal portion of medial lobe of clypeus deeply depressed below adjacent portions of frontal lobes; head as broad as long; longitudinal cephalic rugae (wrinkles) nearly straight or parallel (Fig. 4.28a) .............. 3
    Distal portion of medial lobe of clypeus only slightly depressed below adjacent portions of frontal lobes; head longer than broad; longitudinal cephalic rugae curving toward the eye (Fig. 4.28b) ................ 4

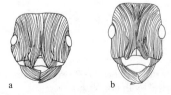

**Figure 4.28.** Face view. (a) *P. barbatus* (b) *P. occidentalis*

3(2) Pronotal rugae very coarse, irregular, widely spaced, and wavy; dorsum of petiolar node with coarse, irregular rugae (Fig. 4.29a) . . . . . . . *P. rugosus*

Pronotal rugae not coarse or wavy; dorsum of petiolar node without coarse, irregular rugae (Fig. 4.29b) . . . . . . . . . . . . . . . . . . . . . . . . . . . . *P. barbatus*

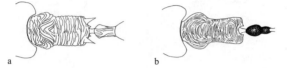

**Figure 4.29.** Dorsal view of mesosoma and node. (a) *P. rugosus* (b) *P. barbatus*

4(2) Cephalic rugae forming concentric loops above eyes . . . . . . . . . . . . . . 5

Cephalic rugae not forming concentric loops above eyes . . . . . . . . . . 6

5(4) Interrugal punctation on head moderate to strong; interrugal spaces sub-opaque (Fig. 4.30a) . . . . . . . . . . . . . . . . . . . . . . . . . . . . . . . *P. maricopa*

Interrugal punctation on head weak or absent; interrugal spaces shiny (Fig. 4.30b) . . . . . . . . . . . . . . . . . . . . . . . . . . . . . . . . . . . . . . . . . *P. californicus*

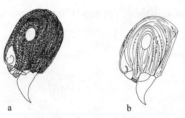

**Figure 4.30.** Profile of head. (a) *P. maricopa* (b) *P. californicus*

6(4) Basal tooth of mandible offset, meeting the short basal mandibular margin at a pronounced angle (Fig. 4.31a) . . . . . . . . . . . . . . . . . . *P. occidentalis*

Basal tooth of mandible not offset, meeting the basal mandibular margin at a straight angle (Fig. 4.31b) . . . . . . . . . . . . . . . . . . . . . . . . . . *P. salinas*

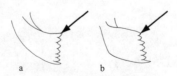

**Figure 4.31.** Face view of mandible. (a) *P. occidentalis* (arrow: offset tooth) (b) *P. salinas* (arrow: basal tooth not offset)

# Harvester Ants

*Messor* Species

## IDENTIFYING CHARACTERISTICS

These ants resemble harvester ants of the genus *Pogonomyrmex* but their stingers are vestigial; they can only bite in defense. Workers range in size from 4 to 7 mm and vary in color from reddish brown to black. The antenna has twelve segments and an indistinct four-segmented club. The epinotum bears conspicuous spines and is depressed well below the level of the pronotum. The mesonotum forms a steeply sloping declivity between the pronotum and the epinotum. The *Messor* species can be differentiated from *Aphaenogaster* species by the appearance of their head, which is quadrate and not narrower behind the eyes than in front of the eyes. Many possess a psammophore.

## DISTRIBUTION

Most of the more than one hundred *Messor* species originated in dry regions of the Northern Hemisphere. Species found in the western United States include the "long-spined harvester ant" (*M. andrei*) and the "jet-black harvester ant" (*M. pergandei*), both of which occur in California, Arizona, Nevada, and into Mexico. *Messor structor,* the only *Messor* species found in Europe, is adapted to arid habitats in central Europe and is found in southwestern Germany, eastern parts of Austria, Hungary, and Slovakia (Czechowski et al. 2002; Seifert 2007).

## BIOLOGY AND HABITS

These desert-dwelling ants harvest enormous quantities of seeds. Their nest entrances have either an irregular disc of gravel or a mound or crater of excavated soil and are surrounded by chaff piles. Some of the largest ant colonies in North America belong to *M. pergandei* (Rissing 1988).

Foraging activity of *M. pergandei* begins in the early morning before sunrise and increases until midmorning, then ceases until late afternoon when it resumes and continues until dusk (Creighton 1953). Foraging columns may be 40 m (131 ft) long with as many as seventeen thousand workers (Wheeler and Rissing 1975). Their walking speed varies with soil temperature from 5 mm/sec at 16 °C to 66 mm/sec at 42 °C (Wheeler and Wheeler 1986). During the hot midday hours the ants work in the nest excavating soil and cleaning out seed husks.

Harvester ants can be belligerent, and their bite has been described as tenacious and annoying (Ebeling 1975). Homeowners may not want to have these ants in their yard because the ants are aggressive and their nests are large and unsightly, but they rarely enter structures.

CONTROL

Messor *species* are considered nuisance pests when they nest near structures. When necessary, nests can be eliminated with an insecticide drench.

### Key to Species of *Messor*

1 Propodeal spines as long as petiole (Fig. 4.32a) . . . . . . . . . . . . *M. andrei*
  Propodeal spines not as long as petiole (Fig. 4.32b) . . . . . . . . . . . . . . 2

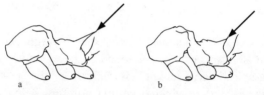

**Figure 4.32.** Profile of mesosoma. (a) *M. andrei* (arrow: long propodeal spine) (b) *M. pergandei* (arrow: short propodeal spine)

2(1) Jet black; propodeal spines short (Fig. 4.32b) . . . . . . . . . . *M. pergandei*
     Propodeum without spines or with blunt tubercles (Fig. 4.33) . . . . . . . . .
     . . . . . . . . . . . . . . . . . . . . . . . . . . . . . . . . . . . . . . . . . . . . . . . *M. structor*

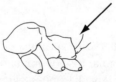

**Figure 4.33.** Profile of mesosoma, *M. structor* (arrow: blunt propodeal spine)

## Fire Ants

*Solenopsis* species (Plate 4b)

IDENTIFYING CHARACTERISTICS

Workers have a two-segmented petiole, a ten-segmented antenna with a two-segmented club, and a stinger. The two major urban pest species in the United

States are the red imported fire ant (*Solenopsis invicta*) and the southern fire ant (*S. xyloni*). Workers of the red imported fire ant are polymorphic, ranging in length from 1.6 to 5 mm, and are almost uniformly dark reddish brown; under direct sunlight they appear almost black. The large workers are called majors, the small workers minors, and those in between medias. The winged reproductives are larger than the workers, although the females are similar in color; the males are black. Southern fire ant workers are bicolored with a yellowish red head and thorax and a dark gaster. Also polymorphic, they range in size from 1.6 to 5.8 mm long. Because of extensive overlap in morphological characters, the two species are difficult to differentiate and may require molecular techniques (Jacobson et al. 2006). Another native fire ant that might be confused with these two species, and especially with *S. xyloni,* is the "desert fire ant" *S. aurea.* It can usually be distinguished by its golden yellow color with pale brown bands on the margin of its gaster (Snelling and George 1979). A fourth species, *S. geminata,* commonly known as the tropical fire ant is also often confused with *S. xyloni* (Smith 1965). However, major workers of this species have large bilobed heads. The black imported fire ant *S. richteri* is predominantly black with yellow mottling (Trager 1991).

## DISTRIBUTION

Most of the approximately 180 species of *Solenopsis* are found in the Neotropical and southern Nearctic regions. The red imported fire ant originated in lowland areas in South America, particularly Argentina and Brazil. It was most likely introduced into the United States between 1933 and 1945 (Lennartz 1973) on cargo ships from South America that docked in Mobile, Alabama. USDA maps show the infestation advancing away from Mobile in concentric circles.

Today, red imported fire ants infest all of the southern states from Florida to eastern Texas, and extend northward into southern Oklahoma, Arkansas, Virginia, and Tennessee. The arid climate of the Southwest has impeded their spread westward, but outbreaks have occurred in desert cities from El Paso in western Texas, through New Mexico and Arizona, and into California. These outbreaks have been associated with commerce, with the ants being transported in trucks, trains, or other vehicles, typically in nursery stock. Red imported fire ants are now established in southern California. There is one recorded occurrence in the state of Washington.

In addition to the United States, the red imported fire ant has been introduced into Australia, China, Mexico, and various islands including New Zealand, Taiwan, Hong Kong, Puerto Rico, the Virgin Islands, and some other Caribbean islands (McGlynn 1999; Holway et al. 2002). Their ongoing global expansion does not yet include Europe, although a woman in Málaga, Spain, suffered an anaphylactic reaction when she was stung by *S. invicta* while handling infested wood from South America (Fernández-Meléndez et al. 2007).

The southern fire ant, native to North America, is found in the southern United States and northern Mexico. In certain parts of its range it is also known as the "California fire ant" and "cotton ant" (Taber 2000). Less abundant but often coexisting with the southern fire ant is the "desert fire ant" (*S. aurea*), whose range is limited to the Colorado Desert in California and north into southern Nevada along the eastern Mojave Desert (Snelling and George 1979). The tropical fire ant (*S. geminata*) has been largely displaced by *S. invicta* in the southeastern United States but has spread to other parts of the world including Australia, southern Africa, India, and many Pacific islands such as Hawaii, Guam, Okinawa, and the Galápagos (McGlynn 1999).

The black imported fire ant was introduced into the United States before *S. invicta* was but has been largely displaced by that species except in the Artesia area of northeastern Mississippi, where populations of *S. richteri* and *S. invicta* x *S. richteri* hybrids still exist (Tschinkel 2006).

## Biology and Habits

Mature colonies of red imported fire ants can have 200,000 to 300,000 workers and can be either monogyne or polygyne. A mature queen can lay hundreds of microscopic eggs each day during peak seasonal activity. The eggs hatch into grublike larvae in 7 to 10 days (Fig. 4.34). After another 1 to 2 weeks these larvae molt into pupae that resemble curled-up, light-colored adults that cannot walk. Over the next 1 to 2 weeks the pupae acquire the reddish brown pigmentation of the adult through a process known as melanization. At the final molt, the pupae become either adult workers or reproductives (Fig. 4.35).

Monogyne colonies have one queen per mound and are territorial. They reproduce by mating flights. After mating, the males die and the newly mated queens seek out nest sites. They are not strong fliers, but with a wind behind them, they may fly for miles before landing. They are attracted to and land on reflective surfaces on pools, parking lots, and truck beds, and in the latter case may be transported for many miles. Typically, however, a newly mated queen lands on the ground, removes her wings, and searches for moist, soft soil where she digs a small hole. Once protected in the soil, she seals the entrance to the hole and begins laying eggs. After the first workers emerge (called minims because of their small size) foraging commences and the incipient colony begins

**Figure 4.34.** Red imported fire ant (*Solenopsis invicta*) workers with reproductive brood

**Figure 4.35.** Red imported fire ant (*Solenopsis invicta*) nest with workers and brood

to grow. In 1 or 2 years the colony matures and produces large numbers of alates in preparation for spring mating flights. Mating flights may also occur at other times when the weather is favorable. The alates prefer to fly after a rain on a warm, clear day with little wind.

Polygyne colonies are polydomous and nonterritorial. Consequently, they can attain higher population levels and mound densities than can monogyne colonies. A polygyne infestation may have hundreds of mounds per acre instead of the thirty to forty mounds per acre typical of monogyne colonies. In addition to mating flights, polygyne colonies can also reproduce by fission or budding (Vargo and Porter 1989), an adaptation that allows them to invade areas where the weather is unfavorable for mating flights.

Red imported fire ants are omnivorous and opportunistic. Although they feed on a wide variety of plants and animals (Lofgren et al. 1975), their primary diet is insects and other small invertebrates (Vinson and Greenberg 1986), including some that are crop pests such as the cotton boll weevil (Sterling 1978), sugar cane borer (Reagan 1981), and tobacco budworm (McDaniel and Sterling 1979, 1982). They also scavenge and feed on carrion.

The red imported fire ant has a particularly bad reputation with respect to its environmental impact. There is no question that in heavily infested areas this species can become a dominant ecological force, sometimes displacing native species of ants and eliminating other invertebrates (Porter and Savignano 1990) and vertebrates (Lofgren 1986). Ground-dwelling animals such as nestlings of quail and other ground nesters are particularly vulnerable to their attack. The disappearance of the horned toad in some areas of Texas is attributed to the fire ant's displacement of harvester ants, which are the lizard's main food (Wojcik et al. 2001). Despite these isolated cases, however, it remains to be determined whether these negative impacts are manifested on a wider scale (Tschinkel 2006).

Researchers at Texas A&M University have determined that red imported fire ants are attracted to electricity. One of the unfortunate consequences of this behavior is that they often infest traffic signal boxes and home air conditioners, where they cause electrical shorts by chewing on the insulation.

Red imported fire ants' notoriety, however, is mainly due to their aggressive behavior and painful sting, which they inflict in unison after crawling en masse onto an unwitting victim. In order to sting, they first grab the skin with their mandibles to gain leverage, and then curl their abdomen to insert the stinger. The venom contains piperidines, which cause a burning sensation, and proteins, which can cause life-threatening anaphylactic shock in a small percentage of the population. The sting causes a characteristic white pustule to form on the skin. Southern and golden fire ant stings do not usually cause such pustules.

**Figure 4.36.** Red imported fire ant (*Solenopsis invicta*) nest mound

**Figure 4.37.** Red imported fire ant (*Solenopsis invicta*) mound opened to show galleries

Although native fire ants can sting, they are less aggressive than their exotic sisters, which attack any object that touches their mound. Nevertheless, the stings of native fire ants have caused medical emergencies. In 2002, a baby died in Phoenix from an anaphylactic reaction to stings of southern fire ants (*S. xyloni*) (Klotz et al. 2005), and there are also reports of allergic reactions to stings by the "desert fire ant" (*S. aurea*) (Hoffman 1997).

The nest mounds of southern fire ants are irregular craters with multiple obscure entrances, whereas the mounds of the red imported fire ant tend to be built up into domes (Figs. 4.36, 4.37). Southern fire ants also nest under objects such as stones, and sometimes in woodwork or masonry (Smith 1965).

Southern fire ants are omnivores with a distinct preference for oily meats and nuts (Snelling and George 1979) and other high-protein foods (Smith 1965). They also chew on clothing, kill young poultry, feed on various plants and seeds, gnaw vegetables and fruits, remove the rubber insulation from around telephone wires, and tend homopterans for honeydew (Smith 1965).

## CONTROL

Total elimination of fire ants from a single property is possible for a limited time, but the ants will reinvade, so regular inspections and treatments are necessary to keep their numbers at a minimum level. Texas A&M University and its extension service have devised the "two-step method" for fire ant control (Merchant and Drees 1992), a simple approach that can effectively control red imported fire ants when applied once or twice a year. Step 1 involves broadcasting a slow-acting granular bait and then leaving the colonies undisturbed for several days to give the foragers the opportunity to carry the bait into the nest and distribute it to the workers, brood, and queens. Broadcast treatments can control colonies even when their mounds are not visible. Step 2 involves applying a fast-acting residual insecticide to the mounds. Formulations for mound treatments include dusts, liquid concentrates, and granules that are watered into the mound. These formulations destroy the remaining ants quickly.

*Baits.* Most fire ant baits consist of insecticides on processed corn grits coated with soybean oil. Fresh baits get the best results, and they should be applied when the ground and grass are dry and no rain is expected or irrigation scheduled for 48 hours. Baits should be broadcast when the workers are foraging outside the nest. In summer, baits are more effective when applied late in the day when it is cooler and ants are more active. Baits can be applied with a hand-held seed spreader set on the smallest aperture. The applicator should make one or more passes over the treatment site at a normal walking speed to apply the label-recommended rate.

Maxforce with hydramethylnon (0.9%) suppressed populations of southern fire ants at a nesting site of endangered California least terns (*Sterna autillarum browni*) (Hooper et al. 1998). The bait was most effective in the spring when alternative food sources for the ants were minimal in this beach habitat. Some ant colonies persisted, however, indicating that the active ingredient was not killing all the queens, and therefore required several reapplications of the bait.

*Dusts.* Insecticidal dusts should be distributed evenly over the mound. Dust particles adhere to the ants as they walk through the treated soil and are carried into the mound. Control of the colony usually results within a few days.

*Mound Drenching.* A number of water-based products are labeled for application to fire ant mounds and are applied as drenches. They must be applied in sufficient quantities to penetrate the entire nest. Amounts required range from a few ounces for small, newly formed mounds to 3.8 to 7.6 l (1–2 gal) for large, mature mounds.

*Granules.* Long-lasting control of imported fire ants (>1 yr) has been achieved with a single broadcast application of granular fipronil (0.0143%) (Greenberg et al. 2003). The active ingredient in other granular insecticides is released after contact with water. These granular formulations should be spread on top of and around the mound, followed by an application of 3.8 to 7.6 l (1–2 gal) of water.

*Nonchemical Option.* Boiling water (ca. 11.4 l [3 gal] per mound) will eliminate approximately 60% of the nests treated but may be hazardous to the person transporting the hot water (Tschinkel and Howard 1980). The objective is to get the hot water deep into the nest. This method works best on cool, sunny days when the ants are in the upper portion of the mound.

### Key to Species of *Solenopsis*

1 Eyes extremely small, 4–6 ommatidia (units of the compound eye); monomorphic (Fig. 4.38) . . . . . . . . . . . . . . . . . . . . . . *Solenopsis molesta*
   Eyes composed of many ommatidia; workers polymorphic . . . . . . . . . . 2

**Figure 4.38.** Profile of *Solenopsis molesta*

2(1)  Uniform golden yellow color ......................... *S. aurea*

Not colored as above ...................................... 3

3(2)  Mandible sharply curved inward and frequently toothless; petiolar node narrow in profile (Figs. 4.39a, 4.40a) ................. *S. geminata*

Mandible toothed, not strongly curved inward; petiolar node not unusually narrow in profile (Figs. 4.39b, 4.40b). ........................ 4

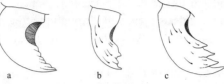

**Figure 4.39.** Face view of mandible. (a) *S. geminata* (b) *S. xyloni* (c) *S. richteri*

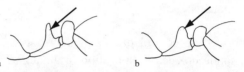

**Figure 4.40.** Profile of petiole and postpetiole. (a) *S. geminata* (arrow: narrow petiolar node) (b) *S. richteri* (arrow: wider petiolar node)

4(3)  Antennal scape short, reaching half the distance between eye and posterior border of head; mandible with 3 distinct teeth (Figs. 4.41a, 4.39b) ...... ................................................... *S. xyloni*

Antennal scape longer, reaching more than half the distance between eye and posterior border of head; mandible with 4 distinct teeth (Figs. 4.41b, 4.39c) ................................................... 5

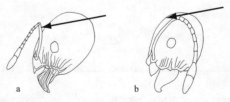

**Figure 4.41.** Profile of head. (a) *S. xyloni* (arrow: short scape) (b) *S. invicta* (arrow: long scape)

5(4)  Black with yellow mottling. ......................... *S. richteri*

Dark reddish brown. ................................. *S. invicta*

# Thief Ants

*Solenopsis molesta* and *S. fugax*

IDENTIFYING CHARACTERISTICS

Workers are monomorphic and extremely small (1.5–2.2 mm long) (Vail et al. 1994). They have a two-segmented petiole and a ten-segmented antenna with a two-segmented club. The body color ranges from yellow or light brown to dark brown (Smith 1965). Thief ants are often confused with pharaoh ants because both are small and similar in color, but the pharaoh ant has a twelve-segmented antenna with a three-segmented club.

DISTRIBUTION

Thief ants are distributed throughout the United States, southern Canada, and northern Mexico (Snelling and George 1979). One species (*Solenopsis fugax*) is native in central Europe, southern England, and the English Channel Islands (Seifert 1996; Czechowski et al. 2002).

BIOLOGY AND HABITS

Thief ants are also known as "grease ants," "sugar ants," and "piss ants" (Thompson 1990). The official common name derives from their habit of nesting next to larger ants and entering their nests to prey on their brood (Wilson 1971). Because they are so often mistaken for other ants, their significance as pests is unknown (Hedges 1997).

Colonies are polygynous, and some mated females return to the nest after the nuptial flight (Snelling and George 1979), which in the United States takes place in late July to early fall (Smith 1965), and in Europe in September and October (Bolton and Collingwood 1975). Queens have been observed carrying a worker on nuptial flights (Smith 1965).

Colonies may contain hundreds to a few thousand ants (Smith 1965). Outdoor nests are constructed in exposed soil, under rocks, or in rotting wood; indoor nests may be in woodwork and masonry, and the ants use the wires in wall voids to travel from one room to another (Smith 1965; Hedges 1992).

Thief ants are omnivores that prey on fire ant queens and other insects (Vail et al. 1994). They occasionally injure sprouting seeds and vegetables and are believed to hollow out seeds for their oil content (Mallis 1969). They also eat carrion and may carry disease-producing organisms into kitchens (Ebeling 1975), and they may be an intermediate host for poultry tapeworms (Smith 1965).

CONTROL

Protein-based baits may be effective given thief ants' preference for high-protein foods (Smith 1965). Dust insecticides are effective against thief ant colonies that are nesting in void spaces and cracks and crevices (Hedges 1998).

## Leaf-Cutting Ants

*Atta* and *Acromyrmex* Species

IDENTIFYING CHARACTERISTICS

Workers are polymorphic and have a two-segmented petiole. Texas leaf-cutting ants (*Atta texana*) are 1.5 to 12 mm long and have an eleven-segmented antenna that lacks a well-defined club (Smith 1965). Workers also have large, flattened mandibles and a spinose body with two pairs of spines on the pronotum (Smith 1965). Their legs are long, and the body is usually dull dark brown or rust brown (Smith 1965). Workers of *Acromyrmex versicolor* resemble *At. texana* but have three pairs of pronotal spines instead of two (Creighton 1950).

DISTRIBUTION

*Atta texana* is found in eastern Texas, western Louisiana, and northeastern Mexico (Smith 1965); *A. versicolor* is found in the Big Bend area of Texas, southern Arizona, southeastern California, and south into Mexico (Creighton 1950; Snelling and George 1979).

BIOLOGY AND HABITS

Leaf-cutting ants cause great economic loss—millions of dollars' worth in many South American countries (Cherrett 1986)—by defoliating trees, shrubs, and crops (Fig. 4.2). Colonies of some species may contain a million or more workers (Cherrett 1986) and can defoliate a tree overnight. In Texas and Louisiana, they are serious pests in pine plantations, where they kill seedlings (Grosman et al. 2002). Also known as cut ants, they use the vegetation as a substrate to grow fungi, which is their source of food. Mature nests contain numerous fungal and brood chambers and have multiple entrances surrounded by crater-shaped mounds. The nests of *A. versicolor* are generally smaller than those of *At. texana* and are usually located in more arid areas (Creighton 1950).

**Figure 4.42.** Leaf-cutting ant (*Atta* spp.) major worker cutting a piece out of a leaf

*Atta texana* colonies are polygyne and may contain more than 100,000 workers (Dash et al. 2005). Their nests are located in well-drained, sandy or loamy soils and may cover an area of 418 m² (4500 ft²). Nests may be up to 6 m (20 ft) deep and have a thousand or more entrances and hundreds of chambers (Smith 1965). A 4- to 6-year-old, medium-sized nest of *At. texana* in a pine plantation in northern Louisiana that was excavated with a bulldozer contained 169 cavities, 97 of which were fungus gardens (Moser 2006). Moser measured vertical tunnels that extended 7.6 m down and speculated that the ants might tunnel as deep as 32 m to reach the water table.

Colonies of *A. versicolor* consist of only a few thousand workers. Nests are generally constructed in sandy regions (Snelling and George 1979) and are often founded by groups of newly mated queens (Fisher and Cover 2007). Dealated females of this species have been observed on foraging trails (Snelling and George 1979).

Texas leaf-cutting ants forage primarily at night in the summer and during the day at other times of the year when the temperature is between 7.2 and 32.2 °C (45–90 °F) (Smith 1965). Foragers follow conspicuous aboveground trails that may extend 183 m (600 ft) from the nest (Smith 1965) as well as underground horizontal tunnels (Moser 2006). Snelling and George (1979) observed *A. versicolor* foraging in spring and fall, but not during hot summer months.

Mating flights of *At. texana* occur at night from early April into June (Smith 1965), with the peak flight period occurring from the last week of April to approximately June 1 (Moser 1967). *Acromyrmex versicolor* mating flights have been reported at dawn in July in Arizona and during November in California (Snelling and George 1979). The female alates carry a fungal pellet from their parent nest to "seed" their fungal garden.

CONTROL

Although primarily an agricultural problem, leaf-cutting ants are considerable pests for homeowners when they damage shrubs, rose bushes, fruit trees, and vegetable gardens (Grosman et al. 2002), and they occasionally invade households to forage on cereals and other foods (Smith 1965; Hedges 1997). Several highly effective baits have been developed, but the organochlorine active ingredients in them are no longer registered for use in the United States. Fumigation with methyl bromide, another effective control measure for leaf-cutting ants, has also been phased out of use. Defatted corn grit baits containing hydramethylnon (1%) and sulfluramid (0.6%) are effective against *At. texana* (Cameron 1990), and experimental citrus pulp baits containing fipronil (0.03%) or sulfluramid (0.5%) look promising (Grosman et al. 2002). Baits should be applied in the spring when the ants begin their activity; small incipient nests can be drenched with a contact insecticide (Dash et al. 2005).

**Key to *Acromyrmex versicolor* and *Atta texana***

Mesosoma with more than 3 pairs of dorsal spines; size small, 6 mm or less (Fig. 4.2a) . . . . . . . . . . . . . . . . . . . . . . . . . . . . . . . . . . . . . . . . . . . . . . . . . . . . . . . . . *A. versicolor*

Mesosoma with 3 pairs of dorsal spines; size varies from 2–12 mm (Fig. 4.43) . . . . . . . . . . . . . . . . . . . . . . . . . . . . . . . . . . . . . . . . . . . . . . . . . . . . . . . . . . . . . *At. texana*

**Figure 4.43.** Profile of *At. texana*

**Acrobat Ants**

*Crematogaster* Species (Plate 4c)

IDENTIFYING CHARACTERISTICS

Workers are monomorphic and have an eleven-segmented antenna with a three-segmented club. They have a pair of spines on the propodeum and a spatulate stinger, which is flexible and permanently exserted (Fisher and Cover 2007). The petiole articulates with the dorsal surface of the heart-shaped gaster, which is flat on top, convex on the ventral surface, and sharply pointed

**Table 4.1.** *Crematogaster* in North America and Europe

| Species | Geographic distribution | Size and color |
|---|---|---|
| *lineolata* | From southern Canada south along the Rocky Mts. and east to FL | 2.5–3.5 mm; body light brown to dark brown to blackish |
| *ashmeadi* | TX east to VA and FL | 2.6–3.2 mm; head and thorax ranging from light reddish brown to brown to black |
| *laeviuscula* | IN to NJ and south to TX and FL | 3.3–4 mm; body varying from yellowish or very light brown to brown to blackish |
| *cerasi* | Southern Canada south through the eastern Rocky Mts. to NM and east through the Black Hills of SD to the Ozark Mts. and GA | 2.6–4 mm; head and thorax varying from reddish brown to dark brown or almost black |
| *californica* | Southern CA and Baja California, Mexico | 2.5–3.5 mm; head and thorax may be reddish and the gaster black |
| *scutellaris* | Mediterranean region of Europe | 2.8–4.0 mm; head reddish, mesosoma and gaster dark brown |

*Sources:* Smith 1965; Seifert, 1996; Snelling and George 1979; Czechowski et al. 2002; Sellenschlo 2002a.

at the posterior end. The common name derives from the scorpion-like posture, with the gaster raised above the head and thorax, that acrobat ants assume when alarmed (Wheeler and Wheeler 1986). Size and color vary with the species (see Table 4.1).

## DISTRIBUTION

Acrobat ants are only occasional household invaders in most of the United States, but in certain regions such as eastern Arkansas and western Tennessee they are important structural pests (Hedges 1997). Smith (1965) recognized four species of *Crematogaster* as household pests in the eastern United States: *C. lineolata, C. ashmeadi, C. cerasi,* and *C. laeviuscula.* Other household pest species include *C. californica* and *C. scutellaris* (see Table 4.1 for their geographic distributions). In Europe, there is one case of *C. scutellaris* being transported from Italy to northern Germany in a camping vehicle and surviving the winter in the insulation (Sellenschlo 2002a), and a colony was observed for several years in the timbers of an unheated shed in northwestern Germany (P. Lieving, pers. comm., 2006).

## BIOLOGY AND HABITS

Acrobat ants nest in a variety of sites, usually in dead wood, including branches and stems of trees and other plants, rotten logs, tree holes, and stumps (Vail et al. 1994). Nests may also be constructed in the ground beneath objects and inside empty nuts and insect galls (Smith 1965). *Crematogaster ashmeadi* and *C. laeviuscula* may build arboreal or semiarboreal nests (Buren 1958). *Crematogaster* colonies may be large with multiple queens (Fisher and Cover 2007).

Some species construct carton brood chambers in their nests and carton sheds over the homopterans they tend (Creighton 1950). All feed on honeydew and both live and dead insects. In households, they feed on sweets and meats (Vail et al. 1994). Workers of some species such as *C. lineolata* and *C. cerasi* are aggressive and may bite and emit a repulsive odor when the colony is disturbed (Smith 1965). The source of the odor appears to be a drop of liquid exuded from the stinger of alarmed ants (Buren 1958).

Structural infestations often originate in a nest outside in a tree, stump, or log. Indoors, they nest in woodwork, particularly doorframes or window frames that have been damaged by insects or moisture (Hedges 1997). A disturbing trend is their infestation of foam-core panel insulation. They have also been known to cause short-circuits by stripping the insulation from wires (Smith 1965).

CONTROL

Treatment of the nest with a residual insecticide is the most effective control method for acrobat ants (Hedges 1998). For indoor infestations, this may entail drilling and injecting an aerosol or dust formulation, or uncovering the nest and treating it with a water-based spray formulation. Repairing water leaks, increasing ventilation, and unclogging gutters may be helpful measures because acrobat ants tend to nest in areas with high moisture. Outdoor nests of colonies that are causing problems indoors may be treated similarly with a residual insecticide.

**Key to Species of *Crematogaster***

1 Red head, dark brown mesosoma and gaster . . . . . . . . . . . *C. scutellaris*
   Color not as above . . . . . . . . . . . . . . . . . . . . . . . . . . . . . . . . . . . . 2

2(1) Antennal scape short, not surpassing posterior border of head (Fig. 4.44a)
   . . . . . . . . . . . . . . . . . . . . . . . . . . . . . . . . . . . . . . . . . . . . . . . . . . . . . 3
   Antennal scape longer, surpassing posterior border of head (Fig. 4.44b)
   . . . . . . . . . . . . . . . . . . . . . . . . . . . . . . . . . . . . . . . . . . . . . . . . . . . . . 4

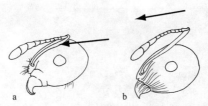

**Figure 4.44.** Profile of head. (a) *C. ashmeadi* (arrow: short scape) (b) *C. lineolata* (arrow: long scape)

3(2) Erect hairs sparse on head and body (Fig. 4.45a) . . . . . . . . *C. ashmeadi*
Suberect hairs on scape and head, long hairs on body (Fig. 4.45b) . . . . . .
. . . . . . . . . . . . . . . . . . . . . . . . . . . . . . . . . . . . . . . . . . . . . . *C. californica*

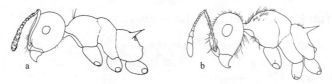

**Figure 4.45.** Profile and mesosoma. (a) *C. ashmeadi* (b) *C. californica*

4 (2) Hairs coarse, bristlelike, well distributed on pro- and mesonotum (Fig.
4.46a) . . . . . . . . . . . . . . . . . . . . . . . . . . . . . . . . . . . . . . . *C. lineolata*
Hairs slender and confined to small area on pronotum (Fig. 4.46b) . . . 5

**Figure 4.46.** Pronotum and mesonotum. (a) *C. lineolata* (b) *C. cerasi* (c) *C. laeviuscula*

5 (4) Mesosoma with few erect hairs (Fig. 4.46b), dorsum of mesosoma with
definite longitudinal striae . . . . . . . . . . . . . . . . . . . . . . . . . . . . . *C. cerasi*
Short, erect hairs on head and mesosoma; dorsum of mesosoma with fine
punctures (Fig. 4.46c) . . . . . . . . . . . . . . . . . . . . . . . . . . . . . *C. laeviuscula*

**Figure 4.47.** European fire ant (*Myrmica rubra*) worker

## "European Fire Ant" and a Related Species

*Myrmica rubra* (Plate 4d) and *M. incompleta*

IDENTIFYING CHARACTERISTICS

Workers are monomorphic and 3.5 to 5 mm in length (Collingwood 1979). The antenna has twelve segments and an indistinct club (Hedlund 2002) (Fig. 4.47). They have a pair of propodeal spines and can inflict a painful sting (Groden et al. 2004), hence the unofficial common name for *M. rubra,* "European fire ant." The body is yellowish brown. Queens are similar in color but a little larger (5.5–7 mm). Males are dark with lighter appendages and are 4.5 to 5.5 mm long (Collingwood 1979). Workers of *M. incompleta* are yellowish red to dark reddish brown with a darker head and gaster (Wheeler and Wheeler 1986).

DISTRIBUTION

There are about one hundred species of *Myrmica,* with *M. rubra* being the most common in Europe. In its native range, extending from Ireland and Great Britain through northern Europe to Siberia, this Palearctic species is not a pest. It was introduced into North America in the early 1900s, most likely on imported plant material, and is now established in parts of the northeastern United States and eastern Canada (Groden et al. 2005). In the 1990s complaints about this ant increased dramatically in the mid-coast region of Maine and on Mount Desert Island, where it has become a major pest (Groden et al. 2005). The range of *M. incompleta* includes the northern states, Canada, and Alaska and extends southward down the Rocky Mountains to New Mexico (Wheeler and Wheeler 1986).

## BIOLOGY AND HABITS

*Myrmica rubra* is found along woodland trails and borders and less frequently in fields and gardens in Europe, where nests are primarily under stones but also in open soil and dead stumps (Wilson 1971). In Maine this species occupies a variety of habitats including lawns and gardens, old fields, scrub-shrub, wet-lands, deciduous forests, and intertidal zones, typically nesting in or under woody debris or leaf litter (Groden et al. 2005). The colonies are polydomous but not unicolonial and have from 297 to more than 10,000 workers and from 0 to 194 queens (Groden et al. 2005). In Europe, a colony may have as many as 20,000 workers and 600 queens. Nests of *M. incompleta* are generally in wet areas around structures and under rocks.

In Europe, mating flights of *M. rubra* occur in late August or September and may be spectacular events when massive numbers of alates congregate over conspicuous landmarks such as high buildings to mate (Wilson 1971). Mating flights are rarely observed in their introduced range, where budding appears to be the primary mode of colony multiplication (Groden et al. 2005). Colonies grow slowly and reach maturity only after about 8 to 10 years (Hölldobler and Wilson 1990). Population densities of *M. rubra* reach higher levels in Maine than in their native range (Groden et al. 2005).

Both species are omnivorous. They tend aphids and coccids for honeydew and also eat small insects.

## CONTROL

Pest management of both species has only recently become an issue in their natural and introduced ranges. Preliminary research indicates that the baits used for imported fire ants (corn grits in soybean oil) are an effective control measure (Groden and Stack 2008).

### Key to Species of *Myrmica*

Frontal carinae curving outward to merge with rugae surrounding antennal sock-ets (Fig. 4.48a) . . . . . . . . . . . . . . . . . . . . . . . . . . . . . . . . . . . . . . . . . . . . . *M. rubra*

Frontal carinae with an angular lobe that is thick and deflected toward the head (Fig. 4.48b) . . . . . . . . . . . . . . . . . . . . . . . . . . . . . . . . . . . . . . . . . . *M. incompleta*

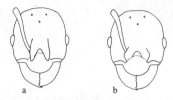

**Figure 4.48.** Face view. (a) *M. rubra* (arrow: curves frontal carina) (b) *M. incompleta* (arrow: angular frontal carina)

## *Aphaenogaster* Species

### IDENTIFYING CHARACTERISTICS

*Aphaenogaster* vary in color (Creighton 1950), and workers are monomorphic or weakly polymorphic (Fisher and Cover 2007) ranging in size from approximately 4 to 6 mm depending on the species. The antenna has twelve segments and an indistinct four-segmented club. The epinotum bears spines and is depressed well below the level of the pronotum. The mesonotum forms a steeply sloping declivity between the pronotum and the epinotum. *Aphaenogaster* workers can be distinguished from *Messor* workers by the shape of the head, which is longer than broad and narrower behind the eyes than in front of the eyes. These ants do not possess a psammophore.

### DISTRIBUTION

*Aphaenogaster* is an ecologically diverse group found throughout North America (Fisher and Cover 2007). *Aphaenogaster fulva, A. lamellidens, A. picea,* and *A. tennesseensis* are found in the eastern United States and are particularly abundant in forests; western species include *A. subterranea valida* and *A. occidentalis,* which are often found nesting in dry, open areas (Creighton 1950).

### BIOLOGY AND HABITS

These ants nest chiefly in soil or rotting stumps and logs, with galleries commonly extending into the soil beneath the wood (Creighton 1950). Some species also nest under leaf litter, tree bark, and stones. Colonies range in size from a few hundred to several thousand individuals.

*Aphaenogaster* are generalist predators and scavengers, feeding on such items as live and dead insects, and seeds (Smith 1965; Fisher and Cover 2007).

They have been reported feeding on cake crumbs and peanut butter but are believed to prefer meat or other high-protein foods (Headley 1949; Smith 1965).

These ants are considered incidental pests for homeowners, particularly when winged forms appear in late summer and fall (Akre and Antonelli 1992), and they have been reported emerging from grooves in floorboards and nesting in the soil around foundations (Smith 1965).

## CONTROL

If control is warranted, a perimeter spray around the exterior foundation is generally sufficient to prevent the ants' entry into a structure.

### Key to Species of *Aphaenogaster*

1  Outer face of frontal lobe with a flange that projects rearward in the form
   of a tooth (Fig. 4.49) . . . . . . . . . . . . . . . . . . . . . . . . . . . . . . *A. lamellidens*
   Outer face of frontal lobe without a toothed flange  . . . . . . . . . . . . . . . 2

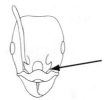

**Figure 4.49.**  Face view of *A. lamellidens* (arrow: flange on frontal lobe)

2(1)  Epinotal spines longer than basal face of epinotum (Fig. 4.50a); postpetiole broader than long . . . . . . . . . . . . . . . . . . . . . . . . . . . *A. tennesseensis*
   Epinotal spines, when present, shorter than basal face of epinotum (Fig. 4.50b); postpetiole as long as broad or longer than broad  . . . . . . . . . . 3

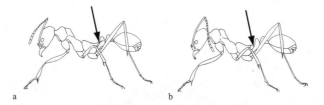

a                                                  b

**Figure 4.50.**  Profile. (a) *A. tennesseensis* (arrow: long spines) (b) *A. picea* (arrow: short spines)

3(2)  Antennal scapes of larger workers surpassing occipital margin by less than length of first two funicular segments (Fig. 4.51a) . . . . . . . . . . . . . . . . . 4

Antennal scapes of all workers surpassing occipital margin by a length greater than the first two funicular segments (Fig. 4.51b) . . . . . . . . . . 5

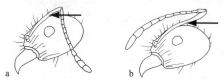

**Figure 4.51.** Profile of head. (a) *A. occidentalis* (arrow: short scape) (b) *A. fulva* (arrow: long scape)

4(3)  Largest worker 6 mm; color castaneous (red to chestnut) brown . . . . . . . .
. . . . . . . . . . . . . . . . . . . . . . . . . . . . . . . . . . . . . . . . . *A. subterranea valida*

Largest worker 4.5 mm; color usually piceous (dark blackish) brown . . . .
. . . . . . . . . . . . . . . . . . . . . . . . . . . . . . . . . . . . . . . . . . . . . . *A. occidental*

5(3)  Anterior border of mesonotum forming a transverse welt; epinotal spines as long as or longer than declivous surface of epinotum (Fig. 4.52a) . . . . .
. . . . . . . . . . . . . . . . . . . . . . . . . . . . . . . . . . . . . . . . . . . . . . . . . . *A. fulva*

Anterior border of mesonotum not forming a welt; epinotal spines not as long as declivous surface of epinotum (Fig. 4.52b) . . . . . . . . . . *A. picea*

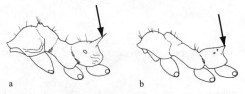

**Figure 4.52.** Profile of mesosoma. (a) *A. fulva* (arrow: long spine) (b) *A. picea* (arrow: short spine)

# CHAPTER FIVE

# Ponerinae

## Subfamily Characteristics

Members of this primitive subfamily have a one-segmented petiole and a well-developed stinger. Most species are inconspicuous and form small colonies in forested areas. They are primarily tropical in distribution and largely carnivorous-predaceous (Snelling and George 1979).

### SCIENTIFIC AND COMMON NAMES

*Hypoponera punctatissima* (Roger, 1859)
*Pachycondyla chinensis* (Emery, 1895): "Asian needle ant"

### Key to Two Species of Ponerinae

Middle and hind tibiae with 2 apical spurs (Fig. 5.1a) . . . . . . . . . . . . . . . . . . .
. . . . . . . . . . . . . . . . . . . . . . . . . . . . . . . . . . . . . . . *Pachycondyla chinensis*

Middle and hind tibiae with 1 apical spur (Fig. 5.1b) . . . . . . . . . . . . . . . . . . .
. . . . . . . . . . . . . . . . . . . . . . . . . . . . . . . . . *Hypoponera punctatissima*

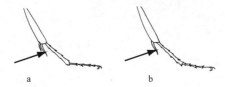

**Figure 5.1.** Hind tibia and tarsus. (a) *Pachycondyla chinensis* (arrow: two apical spurs) (b) *Hypoponera punctatissima* (arrow: one apical spur)

## *Hypoponera punctatissima* (Plate 4e)

### IDENTIFYING CHARACTERISTICS

Workers are 2 to 3 mm long, yellow or brownish yellow, and have a one-segmented petiole with a conspicuously thick node. The body is narrow and elongate, and the head is shiny with very fine punctures. Winged females are 2.7 to 2.9 mm long and slightly broader than workers (Vail et al. 1994). The males are wingless.

### DISTRIBUTION

Members of the genus *Hypoponera* are ubiquitous on a global scale (Wilson 1976). *Hypoponera punctatissima* is a cosmopolitan tramp species that originated in subtropical areas and is widely distributed in Florida (Vail et al. 1994). Several building infestations have been reported in the Northeast (Klotz et al. 2005) and the Pacific Northwest. In central Europe *H. punctatissima* colonies are sometimes found in tropical greenhouses (Seifert 1996; Czechowski et al. 2002). The species has also been transported to Australia, the Arabian Peninsula, the Caribbean, and the Hawaiian Islands (McGlynn 1999).

### BIOLOGY AND HABITS

Colonies of *Hypoponera* typically contain fewer than one hundred individuals and nest in soil or rotten wood (Wheeler and Wheeler 1986). In Florida, nests of *H. punctatissima* are located in leaf litter and soil (Vail et al. 1994), and workers are rarely seen because of their cryptic habits. The alates swarm from outdoor nests in late morning and afternoon during summer, but inside buildings they can swarm from December through July. They are attracted to lights, and sometimes fall on people and sting them (Frishman, pers. comm., 2004). Some people have no reaction to stings; on others, the stings cause red, itchy welts (Vail et al. 1994). One case of respiratory arrest due to an anaphylactic-like reaction to a sting required medical attention (Klotz et al. 2005).

CONTROL

*Hypoponera punctatissima* is considered an occasional pest because the female winged reproductives may sting during their dispersal flights (Vail et al. 1994). Several health care facilities in the Northeast have reported infestations (Klotz et al. 2005). A nursing home in Connecticut was infested by ants—probably brought in on potted plants—that nested in the soil beneath the foundation slab; and employees of a hospital on Roosevelt Island in New York City were stung by ants that emerged from beneath the slab foundation. Winged reproductives were found in two patient rooms in another New York hospital, but no one was stung. Various control strategies were attempted in the Connecticut nursing home. Light traps caught some alates, but residents continued to be stung. Dusting wall voids, attempting to seal cracks in the slab, and a soil termiticide applied as foam also failed. The infestation was finally eliminated by breaking up the floor in the kitchen and inserting chopped mealworms coated with an avermectin dust (Avert) in a rodent bait station. A board placed over the hole in the floor allowed reinspection (Klotz et al. 2006).

## "Asian Needle Ant"

*Pachycondyla chinensis* (Plate 4f)

IDENTIFYING CHARACTERISTICS

Workers are small (ca. 3.5 mm) black ants with light brown mandibles and legs (Japanese database 2003). Dealated females are slightly larger than workers but otherwise similar in appearance (Zungoli et al. 2005). Males are paler and have reduced mandibles (Japanese database 2003).

DISTRIBUTION

It is native to China, but *P. chinensis* is found in other Far East Asian countries such as Korea, Japan, Taiwan, and neighboring islands (McGlynn 1999; Cho et al. 2002). *Pachycondyla chinensis* has also been collected in New Zealand and from Georgia to Virginia in the United States, where it was accidentally introduced in the late 1800s (Hedlund 2003; Zungoli et al. 2005).

BIOLOGY AND HABITS

Colonies of *P. chinensis* are small, ranging in size from one hundred to more than seven hundred workers (Zungoli et al. 2005), and live in moist, rotten wood or in soil beneath objects (Hedlund 2003). In the United States they are found in and around wooded areas and along building foundations (Zungoli et al. 2005); in Korea and Japan they are found in old wooden houses and gardens (Kim and Hong 1992; Fukuzawa et al. 2002). The nests in soil are shallow—approximately 10 cm deep—and those in wood are often associated with termite galleries (Zungoli et al. 2005). Eighteen dealated females were found in one colony with 118 ants (Zungoli et al. 2005). The diet consists of small arthropods (Hedlund 2003).

These ants can inflict painful stings that leave welts that can persist for a week (Zungoli et al. 2005), and allergic reactions have been reported. In ant-infested areas of Korea, for example, the incidence of allergic reactions to stings is 2.1%, much like that reported for imported fire ants in the southern United States (Cho et al. 2002).

CONTROL

Although control methods have not been determined (Cho et al. 2002), preliminary recommendations include nonchemical measures to discourage female alates from entering structures such as installing tight-fitting screens, moving outdoor lights away from structures, and removing mulch or objects that may serve as nesting sites (Paysen et al. 2007). Established nests can be eliminated with an insecticidal drench.

# CHAPTER SIX

# Pseudomyrmecinae

## Subfamily Characteristics

Members of this subfamily are slender, elongate ants with a two-segmented petiole and a well-developed stinger that can administer a potent sting. They typically form small colonies inside preformed plant cavities such as in twigs, thorns, and galls.

### SCIENTIFIC AND COMMON NAMES

*Pseudomyrmex:* Twig ants
    *Pseudomyrmex gracilis* (Fabricius, 1804): "Elongate twig ant"
    *P. ejectus* (F. Smith, 1858)

## Twig Ants

*Pseudomyrmex* species

### IDENTIFYING CHARACTERISTICS

Workers are slender and elongate, and often are not recognized as ants (Wheeler and Wheeler 1973) because of their wasplike appearance and behavior. They have large compound eyes and a two-segmented petiole. Body color varies depending on the species. Workers of the "elongate twig ant" (*P. gracilis*) are 8 to 10 mm long and bicolored orange and black (Vail et al. 1994); those of *P. ejectus* are yellowish brown and 4 to 6 mm long.

## Distribution

*Pseudomyrmex* is primarily a tropical and subtropical genus. *Pseudomyrmex gracilis* is an exotic species that was introduced into Florida and subsequently spread northward into Louisiana and Texas (Hedges 1998; Dash et al. 2005), but *P. ejectus* is a common native species in the South.

## Biology and Habits

Commonly known as twig ants, *Pseudomyrmex* species frequently nest in hollow plant cavities. Most species are arboreal, and some have evolved in close association with plants (Fisher and Cover 2007). *Pseudomyrmex ejectus* nests are generally in hollow twigs on hardwood trees such as live oak. Colonies have multiple nests and queens, and most have fewer than one hundred workers (Klein 1987). *Pseudomyrmex gracilis* nests are constructed in hollow plant cavities of trees, shrubs, grasses, herbs, and vines (Vail et al. 1994). Colonies are small and monogynous (Dash et al. 2005). The workers tend homopterans for honeydew and prey on other insects (Vail et al. 1994; Dash et al. 2005). They are solitary foragers that may be noticed on walls, fences, and railings because of their large size. They are quick and wasplike in their movements and can inflict a painful sting. In Florida, *P. ejectus* is considered an occasional pest because of its sting (Vail et al. 1994). There have been two documented cases of anaphylactic-like reactions to stings: one in Georgia, where ants swarmed out of a gatepost and stung a rancher, and the other in Florida, where ants fell from an oak tree and stung a woman sitting beneath it (Klotz et al. 2005).

## Control

*Pseudomyrmex gracilis* is usually encountered outside around houses and fence rows when plantings become infested (Vail et al. 1994), although ants are occasionally transported indoors on potted plants (Hedges 1998). Control measures are not warranted for this occasional invader (Hedges 1998). Avoiding individual ants is the most effective measure (Dash et al. 2005).

### Key to Two Species of Pseudomyrmecinae

Length 8–10 mm, bicolored orange and black . . . . . . . . . . . . . . . . . *P. gracilis*
Length 4–6 mm, yellowish brown . . . . . . . . . . . . . . . . . . . . . . . . . . . . *P. ejectus*

# CHAPTER SEVEN

# Ecitoninae

## Subfamily Characteristics

Members of this subfamily are called legionary ants or New World army ants. They have a strongly developed stinger and an abdominal pedicel composed of one or two segments (Bolton 1994). The frontal lobes of the head do not cover the antennal sockets, and the antennae attach near the edge of the mouth. The eyes are small or absent (Hölldobler and Wilson 1990).

### SCIENTIFIC AND COMMON NAMES

*Labidus coecus* (Latreille, 1802)
*Neivamyrmex nigrescens* (Cresson, 1872)
*N. opacithorax* (Emery, 1894): Legionary or army ants

### Key to Species of Ecitoninae

1 Tooth present between base and apex of each tarsal claw (Fig. 7.1a) . . . .
. . . . . . . . . . . . . . . . . . . . . . . . . . . . . . . . . . . . . . . . . . . *Labidus coecus*
No tooth between base and apex of each tarsal claw (Fig. 7.1b) . . . . . . 2

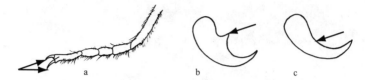

**Figure 7.1.** (a) Tarsus with enlarged tarsal claw (arrow: tarsal claws) (b) *Labidus coecus* (arrow: tooth) (c) *Neivamyrmex opacithorax* (arrow: no tooth)

2(1)  Top border of mandible convex . . . . . . . . . . . . *Neivamyrmex nigrescens*
       Top border of mandible not convex . . . . . . . . . . . . . . . . *N. opacithorax*

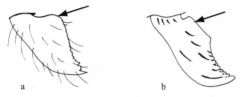

**Figure 7.2.** Right mandible. (a) *Neivamyrmex nigrescens* (arrow: convex border) (b) *N. opacithorax* (arrow: flat border)

## Legionary Ants

*Labidus* and *Neivamyrmex* Species

### Identifying Characteristics

These ants are distinctly polymorphic, with twelve-segmented antennae, a two-segmented pedicel, and eyes that are absent or extremely small. *Labidus* is characterized by having a tarsal claw with a tooth between its base and apex and an extremely short, stout scape that is less than half the length of the head. The body is reddish brown, and the total length is 3 to 10 mm. *Neivamyrmex* species have tarsal claws that lack a tooth and range in size from 2 to 5 mm. The body color is highly variable, and the antennal scape is more than one-half the length of the head (Smith 1965).

### Distribution

*Labidus coecus* ranges from Oklahoma and Arkansas to Texas and Louisiana and south to Argentina (Smith 1965). Members of the genus *Neivamyrmex* occur across the central and southern states, with *N. nigrescens* ranging from Illinois, Iowa, Nebraska, and Virginia south to Florida and west to California and into Mexico (Smith 1965; Wheeler and Wheeler 1986). *Neivamyrmex opaci-*

*thorax* ranges from Kansas to Virginia and south to California, Florida, and into Central America but is less common in the eastern United States than *N. nigrescens* (Smith 1965).

## BIOLOGY AND HABITS

Legionary ant colonies are extremely large; some may contain up to 250,000 workers (Schneirla 1958). The colonies form temporary nests in soil under stones or in logs and stumps when the brood is ready to pupate (Creighton 1950). *Labidus* and *Neivamyrmex* may also nest beneath basement floors and in or around foundation walls (Smith 1965). The members of both genera are subterranean foragers, although *Neivamyrmex* will also make forays above-ground at night (Fisher and Cover 2007). The diet of *Labidus* is high in protein such as dead and live insects, carrion, nuts (Creighton 1950), and small mammals and birds that they attack and kill (Smith 1965). *Neivamyrmex* feeds primarily on other insects (Creighton 1950). Workers of *L. coecus* and *N. nigrescens* may invade houses or other structures in search of meat (Smith 1965). Both species can sting, and *L. coecus* can bite and sting fiercely (Smith 1965). *Labidus coecus* workers have short-circuited telephone wires by removing the lead sheathing (Smith 1965).

Colonies divide by budding, and mating may occur in the nest. Only the males are winged, and they may alarm homeowners when they emerge inside a structure. The same species of *Neivamyrmex* may behave differently in different localities and climates (Wheeler and Wheeler 1986).

## CONTROL

Colonies living in the soil or near foundations can be treated with an insecticidal drench applied in an amount sufficient to penetrate the entire colony. If the colony occurs under a concrete slab, treatment may require drilling through the floor and injecting a residual insecticide that is labeled for subslab treatment of ant colonies (Hedges 1998).

# CHAPTER EIGHT

# Adverse Reactions to Ant Stings and Bites

Defensive behaviors have played an important role in the origin of insect societies. Colonies incapable of defending themselves would have become easy prey to predators seeking to exploit such a concentrated source of food (Schmidt 1986). A colony whose members act together in defense, however, presents a formidable opponent. The current theory holds that ants evolved from nonsocial wasps. By the Cretaceous the primitive wasplike ant *Sphecomyrma freyi* was living in colonies (Engel and Grimaldi 2005) and probably using its sting for defense.

Ants and other stinging Hymenoptera belong to the Aculeata, a monophyletic lineage. The ovipositor of female aculeates has been modified into a stinger, although the stinger is vestigial or absent in many present-day ants, such as those belonging to the subfamilies Formicinae and Dolichoderinae. Ants lacking a stinger generally possess potent defensive secretions, and many can administer a painful bite as well.

Ant venom, a complex mixture of toxic and algogenic chemicals, is a powerful weapon against most vertebrates (Schmidt 1986). Exceptions include horned lizards (*Phrynosoma* spp.) that feed predominantly on harvester ants and have detoxifying factors in their blood that make them relatively insensitive to ant venom (Schmidt et al. 1989), and some species of frogs that eat ants and sequester components of ant venom in their skin for use as defensive chemicals (Eisner et al. 2005).

## Medical Consequences of Ant Stings and Bites

Although human reactions to ant venom vary along a continuum from mild to severe to life-threatening, we all learn to associate stinging ants with pain and injury, and thus to avoid them. For most people, a sting causes only a localized reaction limited to a welt that may cause pain for an hour or two (Greene 2005). Larger local reactions can produce pronounced swelling, as, for example, of an arm or leg after a sting on the finger or toe. Systemic reactions due to toxic envenomations caused by multiple stings or immunologic responses are far more serious. Systemic allergic reactions occur when chemical mediators are released following an immunological reaction that is typically mediated by immunoglobulin E (IgE). For a small percentage of the population, a single sting can precipitate a dangerous, life-threatening allergic reaction called anaphylaxis.

### ANAPHYLAXIS

The term *anaphylaxis,* meaning "without protection," was coined by the French physiologist Charles Richet, who in 1913 received the Nobel Prize in Medicine for his discovery of this phenomenon (Cohen and Zelaya-Quesada 2002). Richet and Paul Portier discovered anaphylaxis while conducting experiments with venom from nematocysts (stinging cells) of the Portuguese man-o-war and sea anemones. They exposed healthy dogs to small doses of venom and then several weeks later repeated the injections. The dogs became ill within seconds of the second injection and died shortly thereafter. Richet and Portier proposed two factors necessary and sufficient to cause this anaphylactic reaction: "increased sensitivity to a poison after previous injection of the same poison, and an incubation period necessary for this state of increased sensitivity to develop" (quoted in Samter 1969).

Current definitions of anaphylaxis reflect our more advanced understanding of its physiological basis; however, the basic tenets of Richet and Portier are still correct today. Much simplified, the process includes a sensitization phase, in which the individual is "set up" for the reaction (Frazier and Brown 1980). Lymphocytes recognize an invading allergen as foreign and become sensitized to it. In the hypersensitive individual, reexposure to the allergen sets in motion a cascade of biochemical events. Lymphocytes that are now sensitized to an allergen release specific IgE antibodies that attach to IgE receptors on mast cells and basophils. The allergen can cross-link these IgE molecules, which alters the cell membrane and leads to the release of histamine and other chemical mediators. Histamine causes contraction of smooth muscle in the large airways and intestines, an increase in vascular permeability, and vasodilation. These events create various manifestations in the affected individual, most commonly urticaria and angioedema—hives and swelling—but the reaction may progress to life-threatening respiratory distress, dizziness, shock, or even death. Gastrointestinal manifestations such as diarrhea may accompany any of the above signs and symptoms.

## TREATMENT

The onset of an allergic reaction typically occurs shortly after exposure to the allergen, so immediate medical attention is critical. Emergency treatment includes administration of epinephrine. In addition, $H_1$ antihistamines like Benadryl™, $H_2$ antihistamines like Pepcid™ or Zantac™, and steroids may also be given to retard progression to a more serious reaction (Pinnas 2001). Antihistamines act by binding to the receptor sites on target cells and thereby blocking them from the effects of histamine. Epinephrine and steroids have multiple anti-inflammatory effects. Individuals known to be sensitive to hymenopteran venom typically carry epinephrine in a small self-injecting applicator (EpiPen).

Allergic reactions to particular allergens are so specific that it is worthwhile to identify the offending insect so that allergists, primary care and emergency physicians, poison control centers, and state entomologists have a record of the insect genus or species that induced the reaction. Immunotherapy is available for the more common causes of sting allergy, such as venoms of yellowjackets, honeybees, and imported fire ants. It involves repeated injections of increasing doses of venom extracts, although in the case of imported fire ants whole body extracts are still used. Immunotherapy may work by (1) activating helper lymphocytes to produce IgG-blocking antibodies that have a high affinity for the allergen, thereby blocking it from binding to mast cells; and (2) producing suppressor T lymphocytes, which suppress IgE production by B lymphocytes.

Unfortunately, no commercial immunotherapy extracts are available for less common causes of sting allergy (Moffitt 2004), although a few physicians specialize in sting allergies and can develop immunotherapy on a case-by-case basis. Hypersensitive individuals should carry an EpiPen and antihistamines so that they can administer an injection of epinephrine at the first sign of anaphylaxis.

## Diagnostics

Toxic reactions to stings can mimic anaphylaxis, but these so-called anaphylactoid reactions have not been shown to be IgE-mediated and can usually be differentiated from anaphylactic reactions by appropriate in vivo skin tests and in vitro radioallergosorbent (RAST) tests. A skin or prick test introduces a small amount of allergen either by intradermal injection, superficial puncture, or scratching (Kuby 1991). If the patient is allergic to the allergen, local mast cells will release histamine and produce a wheal and flare, usually within 20 minutes. In a RAST test the allergen is coupled to a carrier and the patient's serum is added. The amount of IgE bound to the allergen is then measured by adding radiolabeled or enzyme-linked anti-IgE, which binds to the IgE antibody (Kuby 1991).

Using these techniques, allergists and researchers have documented allergic reactions caused by stings and bites of several species of ants that are found in the United States and Europe. What follows is a summary of these cases arranged by subfamily.

### FORMICINAE

Formicine ants do not have a stinger but can spray venom containing formic acid and other nitrogenous constituents into wounds inflicted with their mandibles (Blum and Hermann 1978). Formic acid, the only volatile component thus far detected in their venom, is a potent cytotoxin and also acts as an alarm pheromone; that is, it recruits other workers that also bite the victim (Blum and Hermann 1978). Like members of the other subfamilies, formicines are capable of secreting compounds from the mandibular and Dufour's glands along with venom during an aggressive encounter (Blum and Hermann 1978).

The wood ant (*Formica rufa*) is a widespread species in Europe whose large mounds are familiar sights in northern forests (Hölldobler and Wilson 1994). There is one documented case of a severe anaphylactic reaction to an *F. rufa* bite (Schmid-Grendelmeier et al. 1997). The patient's serum was positive for specific IgE against whole-body extract of *F. rufa*. The glandular source of the allergen, however, was not identified.

The "Florida carpenter ant" (*Camponotus floridanus*), which is also known as the "bulldog ant" because of its pugnacious behavior, is a common household pest in the southeastern United States (Fig. 8.1; Smith 1965). Its venom storage sac occupies most of the volume of its gaster (Eisner et al. 2005). A pos-

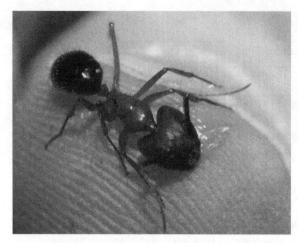

**Figure 8.1.** Florida carpenter ant (*Camponotus floridanus*) major biting into a human finger

sible anaphylactic reaction to its bite was reported in the following case history provided by Dr. Thomas Harper, an allergist in Charleston, South Carolina.

A 6-year-old male with no previous history of adverse reaction to insects other than local reactions sustained a single bite by a large ant on his left thenar eminence (the muscular area at the base of the thumb). Many "Florida carpenter ants" were present on the ground and on a wooden patio table. The bite was very painful, and the insect remained attached to the site of the bite and had to be removed by a family member. Within 30 minutes of the bite, the boy developed facial flushing that spread over his entire body. Generalized urticaria and mild pruritus developed subsequently. His mother noted some mild lip swelling but no coughing, respiratory or swallowing difficulty, or loss of consciousness. He was treated with oral Benadryl™ and was not taken to the emergency room. The area around the bite remained red and swollen for 4 days but subsequently cleared. No pustule formed. The mother was able to obtain several samples of the ants at the time of the bite that the allergist later had identified by an entomologist as *Camponotus floridanus*.

Definitive characterization of this case as anaphylaxis would have required further testing of the child with extracts of *C. floridanus,* which are not commercially available. The positive skin and low positive RAST tests to *S. invicta* indicated that the child had been sensitized previously to imported fire ant stings, which is likely in that region of the country, that there is cross-reactivity between fire ant and carpenter ant allergens, or both.

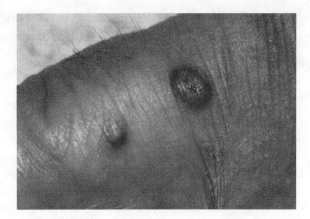

**Figure 8.2.** Characteristic pustules on the skin caused by a sting from a red imported fire ant (*Solenopsis invicta*)

## Myrmicinae

This subfamily contains stinging species with venoms that are typically proteinaceous and highly diverse (Blum and Hermann 1978). Fire ants (*Solenopsis* spp.) are unique because their venoms contain mostly piperidine alkaloids with only trace amounts of protein (ca. 0.1%) (Baer et al. 1979). Piperidines cause histamine release from mast cells (Schmidt 1986) and are responsible for the localized burning sensation typical of fire ant stings. The stings of *S. invicta* and *S. richteri* are the most painful and cause necrotic pustules to form (Fig. 8.2); those of *S. xyloni, S. geminata,* and *S. aurea* are less painful and usually do not form pustules (Schmidt 1986).

The venoms of different fire ant species are highly cross-reactive (Hoffman 1997), meaning that an individual who is sensitized to one species of *Solenopsis* will react to a sting from another species. Allergic cross-reactivity also reportedly occurs in people sensitive to common striped scorpions and imported fire ants, whose geographic distributions overlap (Nugent et al. 2004).

The medical problems caused by imported fire ants are substantial and are increasing in significance with the ants' continued geographic expansion (Kemp et al. 2000). In infested regions of the United States, more than 50% of the population is stung each year (deShazo et al. 1990). A survey conducted in Georgia found that 17% of two hundred randomly selected individuals were sensitized to imported fire ant venom (Caplan et al. 2003), and more than eighty deaths have been attributed to their stings (Kemp et al. 2000). Native species of fire ants cause far fewer problems, although two deaths have been attributed to *S. xyloni.* Both cases were infants less than a year old. In the more recent case in Phoenix, Arizona, the fire ants had invaded a day care facility and were found covering a three-month-old baby girl in her crib. She received hundreds of stings and died as a result of an anaphylactic reaction. Tests for specific IgE for fire ant venom confirmed them as the culprits (Klotz et al. 2004).

Although no deaths have been attributed to *S. geminata,* life-threatening anaphylactic reactions to their stings have been reported (Hoffman 1997). Two U.S. servicemen stationed overseas, one in Guam and the other in Okinawa, had near-fatal allergic reactions to stings from tropical fire ants (*S. geminata*). Both men had been previously sensitized to *S. invicta.* Allergic reactions to "desert fire ants" (*S. aurea*) have been reported in southern California (Hoffman 1997).

Harvester ants (*Pogonomyrmex* spp.) are the other major group of myrmicines whose stings cause anaphylactic reactions. The mammalian toxicity of their venom is higher than that of any other ant species ($<$ 1mg/kg in lethality to mice) (Schmidt 1986). Its constituents include hyaluronidase, a spreading factor; acid phosphatase and phospholipases, important factors in allergic reactions; and a neurotoxic component that is responsible for its lethality to mice (Schmidt 1986). Immunological studies have shown venom cross-reactivity of patients to nine species of harvester ants (Schmidt et al. 1984). The following case describes an anaphylactic reaction to a sting from a rough harvester ant (*P. rugosus*) (Klotz et al. 2005).

A 41-year-old male was brought by ambulance to the University Medical Center in Tucson, Arizona, in September 2003. He stated that he had painful ant "bites" on his groin and that he was short of breath and dizzy. On admission to the emergency room the patient's blood pressure was 89/64 mm Hg; pulse, 112 beats/min; respirations, 22/min; and temperature, 36.0 °C. The man exhibited marked periorbital and perioral edema with cyanosis of the lips; severe wheezing throughout both lung fields; and marked erythema of the right groin area, including the scrotum. Symptoms resolved within 4 hours after treatment with epinephrine, Benadryl™, and steroids. At that time, an extremely tender and enlarged lymph node was present in the right groin along with a large red papule about 1 cm by 1 cm by 1 cm, with piloerection of the surrounding hairs in the vicinity of the papule. The lymphadenopathy, pain, piloerection, and papule at the site of envenomation noted in this patient are characteristic of *Pogonomyrmex* stings. A *P. rugosus* nest was present under the edge of the sidewalk where the patient had been sitting.

Over a 1-year period in Tucson, eight patients were treated for stings by *P. rugosus* and *P. maricopa* (Pinnas et al. 1977); four had systemic allergic reactions and the other four had large local reactions. All developed immediate wheal-and-flare skin reactions to intradermal injections of prepared *Pogonomyrmex* whole-body extract at 1:10,000–1:1,000,000 (w/v) dilutions. A venom preparation was at least one hundred times more potent than the body extract in inducing histamine release.

At least two deaths in Oklahoma have been attributed to stings of *P. bar-batus* (Brett 1950; Young and Howell 1964), and there is one report in South Carolina of anaphylactic reaction to the sting of an unidentified species of *Tetramorium* (Majeski et al. 1974). One possible candidate is the pavement ant (*T. caespitum*), a common urban pest ant that can sting people but has difficulty piercing the skin except in areas where the skin is thin (Gurney 1975).

## PONERINAE

What little is known about the venom of these ants suggests that it is proteinaceous (Blum and Hermann 1978). Anaphylactic reactions have been reported for the stings of two exotic ponerine species in the United States. The more widespread *Hypoponera punctatissima* is considered an occasional pest because the female winged reproductives may sting during their mating flights (Vail et al. 1994). For example, in a nursing home in Connecticut, a nurse developed shortness of breath, dysphonia, and wheezing after being stung and was successfully treated for anaphylaxis (Klotz et al. 2005).

The other ponerine species, *Pachycondyla chinensis,* originated in the Far East and was first noted in the United States in the late 1800s (Zungoli et al. 2005). Two cases of anaphylactic reactions to the sting have been reported in the United States (Leath et al. 2006; Nelder et al. 2006). In some areas of Korea the incidence of systemic allergic reactions (ca. 2.1%) is similar to that for red imported fire ants in the southern United States (Cho et al. 2002). Immunological studies show minimal cross-reactivity of *P. chinensis* to the venom of imported fire ants (Kim et al. 2001).

## PSEUDOMYRMECINAE

The venom of these ants is predominantly proteinaceous but also contains highly pharmacologically active polysaccharides that inactivate human complement and have been used to treat rheumatoid arthritis (Schultz et al. 1978; Schmidt 1982, 1986). There are two documented cases in the United States of anaphylactic reactions to stings of *Pseudomyrmex ejectus* (Klotz et al. 2005). Both cases occurred in the South and involved individuals who had several allergic reactions to stings by these ants. The following case describes one of the incidents.

A rancher in Georgia was stung on the back of the neck by ants that had swarmed out of a gatepost (Klotz et al. 2005). He immediately had difficulty breathing, "turned blue," and became dizzy; his tongue became swollen; and he experienced a feeling of "doom." He drove immediately to a nearby emergency department, where he was successfully treated with epinephrine and steroids.

## CONCLUSION

The summary of anaphylactic reactions to ant stings and bites presented here represents only reported cases, which are probably a small fraction of the actual cases. Many more cases go unreported or are lumped together as venomous stings and bites without specifying the offending organism. In some situations, physicians either do not know how to report them or do not bother to do so. As more cases are reported, the list of ants will inevitably grow to include more species, some of which may be found to be IgE-mediated when evidence for specific IgE antibodies is found. In addition, the incidence of allergic reactions to ants is likely to increase with the accelerating rate of urbanization and globalization. Urban development and sprawl provide disturbed habitats where many species thrive, and increased trade and travel provide entry routes for exotic species. As these ants expand their habitats, more people will encounter them, leading to medical consequences as yet unknown.

# CHAPTER NINE

# Management of Ants

The control of ants in agricultural, natural, and urban settings presents enormous problems to farmers, homeowners, and pest management professionals (PMPs). The sheer diversity of ants and their social behavior, and the ability of some tramp ant species to exploit disturbed habitats have made them worthy adversaries. Broad-scale control efforts began early in the twentieth century, some involving extremely toxic insecticides that we would never think of using today. Various baits incorporating inorganic insecticides such as boric acid, arsenicals, and thallium sulfate were promoted for ant control (Rust 1986). Wall voids, cracks, and crevices were dusted with sodium fluoride powder (Gibson 1916). Mound treatments with calcium cyanide dust were among the first attempts to control the red imported fire ant (Williams et al. 2001).

The presence of the red imported fire ant (*Solenopsis invicta*) in the United States has been the major impetus to developing baits and chemical controls, and the fire ants' spread is now driving the search for biological control agents (see Williams et al. 2001 for a fascinating history of the efforts to control the red imported fire ant). The advent of organic insecticides in the late 1940s spurred a search for new active ingredients and baits. Insecticidal sprays of chlordane, dieldrin, and heptachlor provided long-term control of workers and were widely used in agricultural and urban pest management programs. When the persistent chlorinated hydrocarbons and mirex bait were removed from the market, attention turned to the organophosphate and pyrethroid insecticides and a search for slow-acting bait toxicants. Surface applications of insecticides are often effective for killing workers, but the colony and reproductives frequently escape such treatments (Klotz et al. 1997a). Mound drenches and nest treatments can be effective, but environmental and seasonal factors, inaccessibility of nests, and intensive labor costs affect their success (Banks 1990). The advent of slow-acting toxicants such as fenoxycarb, hydramethylnon, and methoprene in the late 1970s and early 1980s and their success in controlling red imported fire ants led to renewed interest in baits. Recent years have witnessed increased interest in the biological control of red imported fire ants. Experience, however, has taught us that combinations of treatments may be the best strategies for control (Klotz et al. 1997a).

This chapter discusses several major approaches to controlling ants; namely, chemical strategies, habitat and structural modifications, biological control, and integrated pest management (IPM). The primary emphasis is on broad concepts and principles rather than specific recommendations, which are discussed in the species accounts. The distinctive foraging and nesting habits, feeding preferences, and modes of reproduction of the various urban species prevent sweeping generalizations. Some species are widely distributed across North America and Europe while others may have regional importance. In a few instances, several different approaches have been used in a more integrated and progressive approach. This review is intended to serve as a resource to help develop IPM strategies for the many largely ignored urban pest species.

## Chemical Control Strategies

### Baits and Baiting

Baits have long been considered an ideal approach to ant control because they exploit the foraging behavior and social interactions of ants. Baits consist of an active ingredient mixed with food or other attractive substances in liquid or solid form and typically contain less than 5% active ingredient (Bennett et al. 1997). Among the seminal papers dealing with the development of ant baits is Stringer et al. 1964, which outlines four important properties of the active ingredient: (1) delayed toxicity, (2) effectiveness over at least a tenfold range of concentrations, (3) not repellent, and (4) easy formulation with foods and carriers. Ideally, the bait should be attractive to only the target pest ant.

More than seven thousand compounds have been evaluated for use against *S. invicta,* but only six have been developed in commercial baits (Banks et al. 1992), and two of those are no longer marketed. The old definition of delayed toxicity may need some modification, however, to accommodate baits with pro-insecticide (an insecticide that is metabolically activated within the insects body) indoxacarb (Oi and Oi 2006) and other new chemistries being evaluated. Studies with sweet liquid baits have slightly modified the original definition of delayed toxicity to include the time required to kill 50% of the population within 1 to 4 days (Rust et al. 2004) and the concentrations required to kill 50% of the population between days 3 and 8 (Klotz et al. 1997b). The fundamental principle that bait should not impair or inhibit behaviors necessary for trophallaxis (oral exchange of foods and semiochemicals between ants) is common to all of these definitions.

Effective use of baits to control ants requires an understanding of biotic factors such as morphology, nutritional physiology, feeding preferences, and foraging behavior of the species to be controlled. The acceptance, transport, and transfer of active ingredients throughout the colony—and thus their efficacy—are directly affected by these factors.

## MORPHOLOGY OF FEEDING STRUCTURES
## OF ADULT ANTS

Adult ants have chewing mouthparts with mandibles adapted to bite, cut, and rip solid foods or to serve in nest maintenance, defense, and construction and other tasks. Ants also have a number of anatomical adaptations designed to handle primarily liquid foods. The pre-oral cavity enclosed by the mouthparts is the food-receiving pocket. In liquid-feeding insects it is referred to as the cibarium and serves as a pump (Snodgrass 1956). The hypopharynx forms the median postoral lobe of the ventral head. An invagination of the hypopharynx in front of the anterior pharynx forms the infrabuccal cavity (Fig. 9.1). The ridges and hairs that line the infrabuccal cavity remove particles and debris from liquids and compact them into pellets that are later discarded. The infrabuccal cavity of the carpenter ant (*Camponotus pennyslvanicus*) allows only particles less than 100 μm in diameter to pass through to the pharynx. That of the red imported fire ant (*Solenopsis invicta*) filters out particles as small as 0.88 μm (Glancey et al. 1981). When ants feed one another, a behavior known as trophallaxis, they remove many of the larger particles (>10 μm), and this process has been referred to as a social filtration device (Eisner and Happ 1962). Liquids are continually subjected to filtration during trophallaxis.

Liquids filtered by the infrabuccal cavity pass through the buccal tube into the pharynx (Fig. 9.2), the first section of the stomodaeum and differentiated section of the alimentary canal and a second postoral region of suction in liquid-feeding insects. Liquids pass through the pharynx and esophagus en route to the crop (Fig. 9.3). Certain species of the subfamilies Dolichoderinae and Formicinae have a highly modified crop and proventriculus (see Fig. 9.1) to

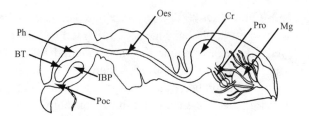

**Figure 9.1.** Digestive system of an ant. *BT,* buccal tube; *Cr,* crop; *IBP,* infrabuccal pocket; *Mg,* midgut; *Oes,* esophagus; *Ph,* pharynx; *Poc,* preoral cavity; *Pro,* proventriculus

facilitate the transport and transfer of this liquid food. The proventriculus is a sclerotized valve between the crop and midgut that regulates food entering the midgut. Only during pumping does liquid pass through the clefts in the valve. Tests of the ability of various species of ants to feed on liquids determined that species with the most modified proventriculus performed the best (Davidson et al. 2004).

The anatomy of the adult ant poses a challenge when formulating baits. Few insecticides are even partly water soluble, and therefore they must be formulated in oils or solvents. If insecticides are suspended in aqueous preparations or solid baits, however, only very tiny particles will be passed during trophallaxis and filtration. Some toxic baits are accepted only by ant species that prefer fats and oils.

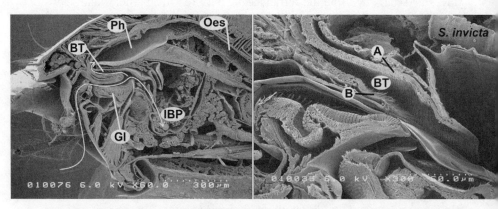

**Figure 9.2.** (a). Sagittal section of the oral area of *S. invicta* worker. (b) Buccal tube lined with filtering setae. *BT,* buccal tube; *Gl,* glossa; *IBP,* infrabuccal pocket; *Oes,* esophagus; *Ph,* pharynx (From Oi 2006, courtesy of the *Journal of Invertebrate Pathology*)

**Figure 9.3.** Argentine ant feeding on dyed sugar water; note greatly expanded abdomen and crop with sugar water

## MORPHOLOGY OF FEEDING STRUCTURES OF LARVAL ANTS

Larvae's ingestion of food particles appears to be limited by the diameter of their esophagus. For example, fourth-instar *S. invicta* larvae cannot ingest particles larger than about 45 μm, which corresponds to the diameter of their esophagus (Glancey et al. 1981). Even proteins are more likely to be fed to the larvae in a liquid form.

## NUTRITIONAL ASPECTS OF FORAGING

Protein is rapidly passed to larvae. Within 48 hours after being offered labeled sugar, *L. humile* larvae had 64% as much sugar as workers on a weight basis, and three times more protein (Markin 1970a). Feeding experiments determined that larvae of *S. invicta* were fed more protein than carbohydrates or oils. Of the total amount of food foraged and distributed by workers, larvae accounted for 30% of the protein, 10% of the oils, and 5% of the sugar water. Only *S. invicta* queens received more protein (Howard and Tschinkel 1981). Liquid lipids and carbohydrates were fed primarily to workers. The high proportion of protein fed to larval red imported fire ants suggests that protein used as a bait base might be very effective in areas with polygynous ants (Weeks et al. 2004). Protein fed in the early spring might be more effective in spreading pathogens or toxicants to developing larvae (Weeks et al. 2004).

Pharaoh ants had to be starved 7 days before workers would forage on peanut oil. The peanut oil was retained by adult workers and also readily distributed to all larval stages, whereas solid egg protein was fed only to older larvae (Haack et al. 1995). Similar patterns of the distribution of oils have been reported for *S. invicta* (Howard and Tschinkel 1981; Weeks et al. 2004).

FORAGING ACTIVITY AND FEEDING PREFERENCES

Foragers typically return to the nest with liquid foods. Less than 1% of the returning foragers of *C. pennsylvanicus* examined by Cannon and Fell (2002), for example, were carrying solid matter in their mandibles. Their crops contained negligible amounts of lipid and glycogen; the major carbohydrate was sugars. Crude protein was a major constituent of the crop, with nitrogenous food making up nearly 50% of the food retrieved. An average of 3.7 Argentine ants per 1000 foragers returned to the colony with solid prey (Markin 1970b). About 70 to 80% of *S. invicta* and *S. geminata* foragers returned with liquid food, most of it of plant origin (Tennant and Porter 1991). Of the solid food, 94% was animal material, and collembolans made up 11.4% of that. Long-legged ant (*Anoplolepis gracilipes*) foragers collected an average 1.2 mg of liquid or 2.8 mg of solid material per trip, with insects making up about 75% of the solid material collected (Haines and Haines 1978).

Foraging preferences and patterns vary greatly depending on the species and the developmental state of the colony. Protein collection by *C. pennsylvanicus* workers, for example, peaked in June and September when mature larvae were present (Tripp et al. 2000; Cannon and Fell 2002). Hydramethylnon baits containing protein were readily accepted at those times and otherwise avoided (Tripp et al. 2000). Peak recruitment to proteinaceous foods by *S. invicta* also occurred during periods of warmer soil temperatures and colony brood production (Stein et al. 1990). Colony development apparently does not affect nutritional preferences of *Monomorium pharaonis, M. floricola,* and *M. destructor* (Eow and Lee 2007).

Seasonal and temperature conditions also have an important influence on daily foraging activities. Table 9.1 summarizes the foraging activities of urban ant species reported in the literature, although the details of foraging behavior of many species have not yet been determined. Certainly, baits applied during the maximum foraging periods are more likely to be taken.

The ability of certain invasive species such as fire ants and Argentine ants to exploit and exclude other native ant species from baits may provide a means to selectively eliminate pestiferous species. Hybrid imported fire ants (*S. invicta* × *S. richteri*), for example, excluded native ants from baits (Gibbons and Simberloff 2005). Similarly, Argentine ants recruited to bait more consistently and in greater numbers than did native species (Human and Gordon 1996). When present in large numbers *L. humile* workers were able to displace *M. minimum* workers from bait stations (Alder and Silverman 2005), but nearly 70% of *L. humile* individuals were killed in one-on-one bouts with little black

**Table 9.1.** Foraging activity and food preferences of urban pest ant species

| Species | Time of day[a] | Temperature range[b] | Food preferences[c] | References |
|---|---|---|---|---|
| *Anoplolepis gracilipes* | B | 21–40 °C | B | Haines and Haines 1978 |
| *Camponotus modoc* | B | (25–30 °C) (8–25 °C) | B | Hansen and Akre 1985 |
| *Camponotus pennsylvanicus* | B | 4–31 °C | P (spring to mid-July) C (late July to early August) | Nuss et al. 2005; Tripp et al. 2000; Fowler and Roberts 1980 |
| *Camponotus vicinus* | N | 2–23 °C | | Bernstein 1979 |
| *Linepithema humile* | B | 15–35 °C | B, C (spring and early summer) C (fall) | Markin 1970b Rust et al. 2000 |
| *Liometopeum occidentale* | E, N | | — | Ebeling 1975 |
| *Monomorium minimum* | D | 32–40 °C | — | Adams and Traniello 1981; Jusino-Atresino and Phillips 1994; Vogt et al. 2004 |
| *Monomorium pharaonis* | B | — | P, O | Eow and Lee 2007 |
| *Monomorium destructor* | | C | | Eow and Lee 2007 |
| *Monomorium floricola* | D | — | O | Eow and Lee 2007 |
| *Paratrechina longicornis* | B | — | C | Meier 1994; Wetterer et al. 1999 |

(continued)

Table 9.1. Continued

| Species | Time of day[a] | Temperature range[b] | Food preferences[c] | References |
|---|---|---|---|---|
| Prenolepis imparis | D (spring) N (summer) | 2–21 °C (10–16 °C) | C | Talbot 1943; Ebeling 1975; Meier 1994 |
| Pogonomyrmex californicus | D | 32–46 °C | seeds, O | Ebeling 1975 |
| Solenopsis invicta | B | 15–43 °C (22–36 °C) | — | Porter and Tschinkel 1987 |
| Solenopsis molesta | D, B | | B | Vogt et al. 2004 |
| Solenopsis xyloni | E, N | 21–43 °C | P | Bernstein 1979; Hooper and 1997 |
| Tetramorium caespitum | | 10–40 °C | — | Brian et al. 1965 |
| Tapinoma sessile | D | 6–35 °C | — | Vogt et al. 2004; Bernstein 1979 |
| Technomyrmex albipes | E, N | 10–32 °C | C | Warner and Scheffrahn 2004; Warner et al. 2004a |
| Wasmannia auropunctata | B | — | C, O | Williams and Whelan 1992; Meier 1994 |

[a]N = nocturnal, D = diurnal, E = evening, B = night and day.
[b]Temperature ranges within parentheses are maximum foraging times.
[c]These are the primary food preferences. C = carbohydrates, P = Proteins, O = oils and lipids, B = both carbohydrates and proteins.

150

ants, whereas only 20% of the little black ants were killed. *Wasmannia au-ropunctata* effectively displaced *Paratrechina longicornis* from baits (Meier 1994). Competition occurs within species as well. Polygynous forms of *S. invicta,* for example, recruited to baits in higher numbers than did monogynous forms (MacKay et al. 1994).

## ACTIVE INGREDIENTS IN BAITS

Effective toxicants must have delayed toxicity and be nonrepellent. Solubility in oils or water is important in formulating them into baits and determining whether ants will accept them.

The insect growth regulators (IGRs) that have been incorporated into baits include methoprene, fenoxycarb, and pyriproxyfen. Effective IGRs prevent worker replacement by killing immatures, causing degeneration of the queen's reproductive organs, shifting caste differentiation from workers to sexual forms, or some combination of these (Williams et al. 1997). The decline in brood levels may be attributable to direct toxicity, disruption of development, and reduction or cessation of egg production (Williams and Vail 1993). Since IGRs do not kill adult workers quickly, they are more likely to be spread throughout a colony. This is especially important when control efforts are aimed at polygynous and polydomous species such as *M. pharaonis* (Oi et al. 2000).

Avermectin B1 (abamectin) is a natural product derived from soil bacteria that acts as a gamma-aminobutyric acid (GABA) agonist. It eliminates the GABA-mediated inhibitory postsynaptic potentials, causing nerve hyperexcitation that results in tremors and uncoordinated movements (Valles and Koehler 2003). Avermectin is a highly effective inhibitor of reproduction in queen red imported fire ants, but it does not rapidly eliminate colonies (Lofgren and Williams 1982).

Boric acid is an inorganic compound that has been used as a dust and bait ingredient for urban pest control for more than a century. Its mode of action is not well understood, but cockroaches and ants fed low concentrations show gross abnormalities and disruption of the cells that line the midgut (Klotz et al. 2002; Habes et al. 2006). Ingestion may also cause neurotoxic action by reducing the activity of acetylcholinesterase (Habes et al. 2006). Boric acid has been a popular active ingredient in aqueous sweet baits because of its moderate solubility in water (5.5% at 25 °C; Merck 1976).

Fipronil, one of the phenylpyrazole class of insecticides, has low water solubility (2.4mg/L at 20 °C; Anonymous 1996) and acts on insects by blocking the GABA-gated chloride channels. It is extremely active against ants and is readily transferred from one ant to another (Soeprono and Rust 2004a, 2004b). In addition to baits, it has been formulated into granules and sprays.

Hydramethylnon, one of the amidinohydrazone insecticides, is a metabolic inhibitor that interferes with energy production (Hollinghaus 1987). In a granular insect bait base, it rapidly kills *L. humile* workers and does not act as a delayed toxicant (Knight and Rust 1991). When formulated in oil, however, it has delayed toxicity.

Indoxacarb is a proinsecticide (i.e., it is bioactivated inside the insect by enzymes to form a toxic metabolite) belonging to the oxadiazine class. It inhibits voltage-gated sodium channels (pyrethroid insecticides, in contrast, prolong their opening) (Zhao et al. 2005).

Spinosad is a macrocyclic lactone produced by soil actinomycetes that acts on the nervous system of insects through the nicotinic acid receptors. The continuous activation of motor neurons causes the insect to die from exhaustion (Salgado et al. 1998).

Sulfluramid belongs to the halogenated alkyl sulphonamide chemical class. It is potentiated by enzymes in the insect's body, and the toxic metabolites inhibit energy production (Valles and Koehler 2003).

The nicotinoid compound imidacloprid was formulated into a bait to control ants with a preference for honeydew. A sucrose-water bait with 50 ppm (0.005%) imidacloprid reduced numbers of *Tapinoma sessile* by more than 80% after 6 weeks, a result similar to that obtained with a boric acid–containing liquid formulation (Higgins et al. 2002). A sucrose-water bait with 0.001% imidacloprid was effective against Argentine ants (Rust et al. 2002). A liquid ant bait with 0.03% imidacloprid provided good control of *Monomorium pharaonis, Lasius niger,* and *Tapinoma melanocephalum* (Pospischil 2008).

## PHYSICAL PROPERTIES OF BAITS

The physical properties of a liquid bait, such as density and viscosity, may influence its acceptance (O'Brien and Hooper-Bui 2005). The rate of intake of sucrose liquids by *Camponotus mus,* for example, is constant up to 30% but declines afterward (Paul and Roces 2003). Sucking yields a higher energy uptake at lower sucrose concentrations, whereas licking seems to be more advantageous at higher sugar concentrations, which have a higher viscosity. Maximum crop filling for *C. mus* occurred with 42.6% sucrose solutions (Josens et al. 1998). Workers will recruit to either low or high sugar concentrations, but feeding time increases as the concentration increases. Ants may not completely fill the crop if they are feeding on dense liquid baits as they would with sugar water. *Linepithema humile,* for example, is not efficient at extracting sucrose from gels and in laboratory experiments consumed twice as much liquid sucrose as gels in 5 days. Mortality was thus significantly higher on sucrose fipronil liquid baits (Silverman and Roulston 2001). Similarly, *S. invicta* workers spend more time feeding on oils than on sugars, and their consumption of sucrose solutions is relatively low regardless of the worker's size (Howard and Tschinkel 1981).

Head width of foragers is directly correlated with the size of bait particles preferred, with the exception of harvester ants that have psammophores (seed beards) (Hooper-Bui et al. 2002). When species have polymorphic workers, the range of particles sizes accepted will vary greatly depending on worker size. *Solenopsis xyloni* workers, for example, remove particles ranging in size from 420 to 2000 μm (Hooper and Rust 1997). Harvester ants pack the psammophore with the smallest particles. Furman and Gold (2006a) found that 2 mm per granule formulated with indoxacarb (about 600 granules/g) provided the greatest activity against red imported fire ants. In general, the ants were inclined to make many visits to smaller particles.

## Bait Timing and Application

Baits placed around the perimeter of structures have been successful in reducing Argentine ant numbers (Forschler and Evans 1994). The purpose of such "diversionary baiting" is to draw ants away from the structures to be protected. Likewise, placing discrete bait stations outdoors on the exterior walls of buildings controlled pharaoh ants before they could get indoors (Oi et al. 1994, 1996; Vail et al. 1996).

Baits containing hydramethylnon may not be stable when exposed to light and should be applied as close to dusk as possible (Vander Meer et al. 1982). Applications of hydramethylnon bait in late afternoon, just prior to the maximum foraging activity of *S. xyloni,* provided maximum efficacy (Hooper et al. 1998), but the efficacy of bait applications against the native ant *M. minimum* did not differ when applied at midday and in the evening (Vogt et al. 2005). Bait treatments for red imported fire ants will have the greatest impact on active ants in the late afternoon in summer, when *M. minimum, T. sessile,* and *S. molesta* are most active (Vogt et al. 2004). A red imported fire ant bait with indoxacarb controlled the non-target species, *M. pharaonis* and *Dorymyrmex pyramicus,* for 3 weeks and *Pogonomyrmex barbatus* for 7 weeks (Furman and Gold 2006b).

Baits stored in garages and storage facilities should be placed away from possible contaminants. Red imported fire ant baits, for example, absorb volatile materials such as insecticides and gasoline, which can affect their acceptance (Benson et al. 2003).

Studies with *L. humile* found no advantage to scattering baits compared with discrete granular bait placements (Silverman and Roulston 2003). More scattered bait was returned to the colony within the first 2 hours, but after that there was no difference in the amounts returned to the nest. Discrete bait placement was also effective in controlling *S. xyloni* (Hooper et al. 1998). Findings such as these have potentially important implications in helping to reduce the runoff of insecticides into urban watersheds. Placement of baits in stations should also reduce the likelihood that nontarget organisms will contact the bait.

BAIT SWITCHING

Foragers of a number of species will switch from carbohydrates to proteins according to the developmental state of the colony or to previous feeding regimes. *Monomorium pharaonis* fed on a carbohydrate-rich diet or a protein-rich diet readily switched to the other bait when given a choice (Edwards and Abraham 1990), sometimes within as little as 2 weeks. Foragers fed two preferred foods, honey and peanut butter, alternated feeding preferences between carbohydrate baits and protein baits. Eight species in the genera *Pheidole, Pogonomyrmex, Dorymyrmex, Formica,* and *Forelius* with easy access to carbohydrates preferred protein, and those with easy access to protein preferred carbohydrates (Kay 2004). These studies support the notion of prebaiting to determine the most acceptable baits prior to initiating large-scale baiting programs.

PERIMETER SPRAYS AND BARRIERS

With the advent of the chlorinated hydrocarbons in the late 1940s, residual sprays and dusts of chlordane, dieldrin, aldrin, and heptachlor largely replaced ant baits and powders for ant control (Mallis 1969). When chlorinated hydrocarbons were removed from the market in the 1980s, and organophosphates such as chlorpyrifos and diazinon in the mid-1990s, pyrethroids such as bifenthrin, cypermethrin, cyfluthrin, lambda-cyhalothrin, and permethrin became the common sprays and granulated insecticides (Rust and Knight 1990). Now that organophosphates are no longer available, pyrethroids and new synthetic compounds such as pyrrole, chlorfenapyr, phenylpyrazole, and fipronil are being widely applied.

Broadcast applications of sprays or granulated insecticides were once considered to act as barriers preventing ants' access to structures either by killing them, by repelling them, or both. Laboratory studies using artificial ant trails determined the toxicity and potential for repellency for a large number of insecticides (Knight and Rust 1990b). Recent research suggests that fast-acting insecticides such as pyrethroids actually prevent ants from establishing trails across treated barriers (Pranschke et al. 2003; Soeprono and Rust 2004b). The treatments give the impression that they are highly repellent when in fact they rapidly kill ants that enter the barrier, preventing them from organizing trails, a phenomenon known as "virtual repellency." Consequently, the application of fast-acting barriers can actually trap ants within barrier treatments and structures (Rust et al. 1996; Gulmahamad 1997).

Factors such as heavy irrigation, dense ground cover, exposure to direct sunlight, porous alkaline stucco and concrete surfaces, and high temperatures interfere with the performance of barrier applications. Applications of nonrepellent insecticides such as chlorfenapyr and fipronil to nonporous substrates such as ceramic tile, however, were effective against *M. pharanois* (Buczkowski et al. 2005). The insecticidal activity of most outdoor barriers for Argentine ants persists about 30 days (Rust et al. 1996). Applying insecticides along structural guidelines such as the lower edges of siding and around window and door frames and eaves, out of direct sunlight, will maximize their residual activity (Tripp et al. 2000).

Fipronil barriers provided excellent reductions of a number of species for 4 weeks or longer (Vega and Rust 2003; Scharf et al. 2004; Klotz et al. 2007). The effectiveness of these barrier applications may be attributable to the slow-acting toxicity of fipronil and its horizontal transfer to nestmates (Soeprono and Rust 2004a). Workers continue foraging and interacting with nestmates for several hours after exposure. Other workers remove the dead workers to graveyards (necrophoresis) and in the process acquire a lethal dose of insecticide. This in part explains the activity of limited or spot treatments against *L. humile* (Klotz et al. 2007) and carpenter ants.

The use of foliar sprays or systemic insecticides to reduce the number of homopteran honeydew sources has been proposed as a means of controlling ants around structures. Quantifying the impact of such treatments against *L. humile* has been difficult (Rust et al. 2003). Treating vegetation with systemic insecticides is recommended for the control of white-footed ants (Warner et al. 2006).

Barriers of imidacloprid and cyfluthrin provided control for 4 and 2 weeks, respectively (Scharf et al. 2004), of *Prenolepis imparis, T. sessile, S. molesta, Crematogaster ashmeadi, Paratrechina longicornis, Formica* spp., and *Camponotus pennyslvanicus.* Odorous house ants were not present at the beginning of the study but were present at treated locations 1 to 8 weeks after the treatment was applied.

Applications of granular insecticides alone around structures provided marginal control of Argentine ants (Rust and Knight 1990; Rust et al. 2003) and

pharaoh ants (Oi et al. 1996), but prevented red imported fire ants from cross-
ing barriers for up to 15 weeks (Pranschke et al. 2003). Their application in
dense ground cover and around areas likely to harbor nests in conjunction with
perimeter sprays has been more effective against Argentine ants (Rust et al.
1996; Rust et al. 2003; Klotz et al. 2007). Fipronil granules scattered in grassy
fields significantly reduced the number of *S. invicta* workers for 65 days
(Loftin et al. 2003); applied on golf course fairways, fipronil provided 97% reduction at 9 months (Greenberg et al. 2003). Applications of bifenthrin granules in simulated nursery situations eliminated Argentine ant foraging for 12
weeks (Costa et al. 2001). Granular insecticides should not be used in locations susceptible to water runoff, such as on driveways, sidewalks, and street
curbs.

The incorporation of granular bifenthrin into potting soil mixes is recommended to prevent ants, especially *S. invicta* and *L. humile,* from establishing
colonies in potted plants. A less toxic alternative might be the use of 2% mint
oil granules (Appel et al. 2004).

MOUND TREATMENTS

Treatment of individual nests and mounds is effective but extremely labor intensive, and the mounds or nests may be on other property or otherwise inaccessible to treatment. Spot applications of chlorpyrifos to mounds of red
imported fire ants provided 80% reductions at week 8, but the colonies simply
relocated beyond the treated area. No insecticide tested killed 100%. Mound
treatments should be considered supplemental treatments to baiting (Collins
and Callcott 1995). Citrus oil formulations provide an alternative mound treatment with activity similar to diazinon against red imported fire ants (Vogt et
al. 2002). Barr (2003) reported that indoxacarb bait provided rapid activity
when applied to mounds of *S. invicta.* Because of its speed of activity it may
eventually replace contact materials such as bifenthrin.

## Habitat and Structural Modification

Modifying the urban environment can alter potential nesting sites, access to
structures, and conditions conducive to pest species. In the early 1900s, ant
traps were wooden boxes filled with grasses and organic matter designed to
entice Argentine ants to establish colonies that were then destroyed by fumi-
gating with carbon bisulphide (Newell 1909). Ants such as *L. humile* and *T.
sessile* will readily nest in pine straw, bark mulch, and other ground covers

used around structures. Aromatic cedar mulch killed *L. humile* and *T. sessile* colonies, and *L. humile* colonies avoided nesting in it for 4 months (Meissner and Silverman 2001). Aromatic cedar mulch applied around trees reduced *L. humile* activity and nests (Meissner and Silverman 2003). However, attempts to integrate the use of repellent mulches and spot insecticide treatments have had limited success (Silverman et al. 2006). Finely ground red cedar mulches were somewhat repellent to *S. invicta,* but most of the hardwood, pine, and cypress mulches tested were not deterrent (Thorvilson and Rudd 2001).

Trimming vegetation away from structures eliminates potential nesting sites and structural guidelines that lead ants into buildings. Among the practical measures Hedges (1997) recommended are: repairing damaged trees and sealing off tree holes; trimming all trees away from the walls and roof; removing piles of lumber, wood, bricks, and other materials; and keeping landscape mulch to a thickness of 5 cm or less. Very fine particles can serve as a repellent to ant foraging; for example, barriers of fine sand (<230 mesh; i.e., sand fine enough to pass through a 230-mesh screen) prevented *L. humile* from foraging (Rust et al. 2002). California campers frequently protect their recreation vehicles by surrounding the tires with powdered household cleanser. Compounds such as 2-nonanol, 2-methyl hexanoic acid, octanoic acid, and 1-decanol repel *S. invicta,* and octanoic acid has prevented ants from moving into potted plants (Vander Meer et al. 1993).

Pest control can begin at the time of construction. For example, wall voids can be treated with insecticidal dusts before insulation is added to eliminate access to ants such as acrobat ants, carpenter ants, odorous house ant, and velvety tree ants. If dusts remain dry, they will remain active for years (Hansen and Klotz 2005).

## Biological Control

Exploration for biological control agents has become increasingly important in recent years, especially for invasive species such as *S. invicta* and *L. humile.* Biological control of urban ants has specific problems, including (1) residents' unwillingness to coexist with any pest insects, (2) lack of proven biological agents, (3) interactions between indoor and outdoor ants, and (4) manufacturers' inability to formulate biological agents into suitable products (Pereira and Stimac 1997). Williams et al. 2003 and Williams and deShazo 2004 offer excellent reviews of potential biological control agents for *S. invicta.*

At this point, the only entomopathogenic nematodes (EPNs) known to be effective against social insects are *Steinernema* spp. and *Heterorhabditis* spp.

(Peters 1996). Their usefulness against ants, an exceedingly diverse group, has yet to be determined (Gouge 2005). There are no reports of nematodes belonging to the genera *Steinernema, Neosteinerema,* or *Heterorhabditis* occurring naturally in ants (Peters 1996). Their spread requires direct contact, and they do not spread normally to other fire ant colonies. Self-sustaining biological control agents would be far more efficient (Williams et al. 2003).

Effective delivery of microbial agents to ant colonies can be a significant obstacle to their use in control programs, although *S. invicta* workers readily took mycelia of *Beauveria bassiana* encapsulated in oil-coated alginate pellets (Bextine and Thorvilson 2002). The use of alginated pellets to deliver microbial agents deserves further study.

A number of potentially interesting biological agents are being evaluated for use against the red imported fire ant; namely, the exotic decapitating fly *Pseudacteon tricuspis* and the exotic microsporidium *Thelohania solenopsae. Pseudacteon tricuspis* has been released in eleven states and has become established in seven, and is spreading about 6.2 km per year (Williams and deShazo 2004; USDA Agricultural Research Service 2005). Red imported fire ants learn to avoid the flies and thus forage less. Three attractive features of decapitating flies are: (1) they are specific to red imported fire ants; (2) they can function over seasons, geography, and climate; and (3) they affect ant behavior (Williams and deShazo 2004). Early field evaluations have yet to reveal any impacts of *P. tricuspis* on *S. invicta* populations in Florida (Morrison and Porter 2005), which vary tremendously in densities because of environmental factors such as rainfall and disturbance. *Pseudacteon* parasitoids did not attack *L. humile,* but did attack a closely related *Linepithema* species (Orr et al. 2001). Other flies that do attack pestiferous species may be identified.

Infection of *S. invicta* by the microsporidium *Thelohania solenopsae* causes a chronic disease that debilitates the queen and results in the slow death of the colony. There is evidence that queens transmit the infection to their offspring. Current efforts involve inoculating wild colonies with infected brood. Polygynous forms of *S. invicta* may predominate because colonies with multiple queens do not succumb as quickly as those with monogyne queens. In studies with hydramethylon, red imported fire ant colonies infected with *T. solenopsae* were 2.4-fold more susceptible to hydramethylnon, and field colonies declined much faster after applications of hydramethylnon bait (Valles and Pereira 2003).

Williams et al. (2003) cited the potential of *T. solenopsae* to "(1) be a long-term, environmentally compatible fire ant control tactic that is applicable to a broad range of systems where fire ant controls are not available; (2) reduce reliance on pesticides by slowing reinfestations; (3) protect and conserve ecosystem quality and diversity by reducing fire ant dominance, thereby encouraging the establishment of native ants and other arthropods; (4) be used as a stress factor increasing the susceptibility of *S. invicta* to other pathogens and natural enemies; and (5) be used in an integrated pest management program where infected fire ants would be more susceptible to pesticides."

Another microsporidium, *Vairimorph invictae,* has been successfully introduced into red imported fire ant colonies. Like *T. solenopsae* it hinders colony growth (Oi et al. 2005). The search for other biological control agents that will be effective against pest ant species continues. For example, a number of wasps belonging to the genus *Orasema* reported to be parasitoids of ant pupae may be potential biological control agents (Heraty 1994). Species of ants they parasitized included *W. auropunctata* (Wetterer and Porter 2003), *S. invicta, S. molesta,* and *P. megacephala* (Heraty 1994).

Clearly, exploration for new pathogenic agents and parasitoids for all of the important invasive or tramp ant species is crucial. To date, the efforts have been primarily limited to biological control agents of the red imported fire ant.

## Integrated Pest Management (IPM)

An IPM program for ant control that is adaptable to almost any urban ant pest problem involves four basic steps: (1) inspection and identification, (2) monitoring, (3) prescription treatment, and (4) monitoring (Forschler 1997; Hansen and Klotz 2005). Hansen and Klotz 2005 includes an excellent list of questions for the homeowner regarding carpenter ants that could be easily modified for any pest species in a given area. The list will help to focus both PMPs and homeowners on conditions conducive to ant problems and noninsecticidal solutions (Table 9.2). An educational component for tenants, property owners, and the public can be extremely important for the success of urban IPM programs, particularly when cooperation is required to conduct areawide programs or when biological control agents and chemical treatments may take considerable time to achieve their full impact.

Considerably more effort has been focused on developing broad-based programs to control the red imported fire ant than on any other species. Much of this effort still relies on applications of baits, but more recent efforts include biological agents. Three principal site-specific IPM options include the "two-step method," individual mound treatments, and the "ant elimination method" (Drees and Gold 2003). The two-step method involves broadcast application of bait followed by a mound treatment. Ant elimination typically involves applying long-term residuals over large areas. Combinations of less persistent

**Table 9.2.** Interview questions asked to homeowners for the purpose of obtaining the history of an ant infestation

Interior infestation

1. How long have you seen ants in the house?
2. In what rooms are you seeing the ants?
3. Are the ants winged?
4. How many ants are you seeing in a 24-hour period?
5. Have the ants been attracted to a particular food product?
6. Have you noticed sawdust anywhere that may have been excavated by the ants?
7. Have you heard any rustling noises in the walls?
8. Have you moved potted plants in the structure recently?

Exterior infestation

1. Have you observed ants outside the structure?
2. Do you have pets and feed them outdoors?
3. Have you noticed ant trails along foundations or landscaping timbers?
4. Do you use decorative bark in your landscaping? If so, how deep is the bark?
5. Are trees, shrubs, or plants touching the roof or sides of the house?
6. Have any trees been removed from the yard? If so, were the stumps removed?
7. Are landscaping timbers or railroad ties used in your yard?
8. Have major disturbances such as new water, sewer, or electrical lines occurred?
9. Are there trash receptacles near the structure?
10. Are there plants infested with aphids, scales, or mealybugs around the structure?

*Source:* Modified from Hansen and Klotz 2005.

sprays such as fipronil or imidacloprid and baiting have been effective against *T. sessile* (Scharf et al. 2004). A combination of eliminating honeydew and food sources, chemical treatments, and limiting the ants' access to the structure has been adopted against *Lasius neglectus* (Rey and Espadaler 2004). Chemical treatments include tree fogging, tree trunk spraying, house perimeter injection, and in-house baiting.

A combination of biological control agents and conventional baiting strategies shows promise for the control of imported fire ants. For example, *Pseudacteon curvatus* and *Thelohania solenopsae* have been released in combination with bait applications in large areas of Mississippi (Vogt et al. 2003). One outcome of such areawide treatment programs is the prospect of increasing the native diversity of ants and enhancing their ability to compete with invasive species such as the Argentine ant, red imported fire ant, and little fire ant. In several urban neighborhoods in Texas, hydramethylnon bait reduced *S. invicta* numbers and increased the ant diversity within 2 years. Four or five new ant species were found after the treatments (Riggs et al. 2002).

CONCLUSION

No single approach will be effective in controlling urban pest ants. An effective IPM approach should incorporate both habitat modification and combination treatments. Invasive species will continue to present a major problem to urban communities worldwide, and areawide programs may be necessary to control them. Additional research on the vast majority of species that inhabit the urban environment may help us solve some of the problems presented by urban ant pests.

**APPENDIX 1**

**Scientific and common names of urban ants**

| | Common names | | | |
|---|---|---|---|---|
| Subfamily/Species | English | German | French | Spanish |
| Formicinae | | | | |
| *Camponotus* spp. | Carpenter ants | Rossameisen | Fourmis charpentières | Hormigas del carpinteros |
| *Camponotus herculeanus* | Boreal carpenter ant, Northern carpenter ant | Rossameise | Fourmi rouge charpentière, Fourmi bicolore | Hormiga del carpintero |
| *Camponotus pennsylvanicus* | Black carpenter ant | — | Fourmi noire gâte-bois | — |
| *Paratrechina* spp. | Crazy ants | — | "Fourmis folles" | Hormigas locas |
| *Prenolepis imparis* | Small or False honey ant | — | — | Hormiga falsa mielera |
| *Lasius flavus* | Yellow turf ant | Gelbe Wiesenameise | Fourmi jaune | — |
| *Lasius niger* | Black garden ant | Schwarzgraue Wegameise | Fourmi noir des prés | Hormiga negra común |
| *Lasius brunneus* | Brown ant | Rotrückige Holzameise, Braune Wegameise | — | — |
| *Lasius umbratus* | — | Gelbe Schattenameise | — | — |
| *Lasius fuliginosus* | Yet ant | Glänzendschwarze Holzameise | Fourmi noir des bois | — |
| *Lasius alienus* | Cornfield or moisture ant | Fremde Wegameise | Fourmi noir des pelouses | Hormiga del maizal |
| Dolichoderinae | | | | |
| *Linepithema humile* | Argentine ant | Argentinische Ameise | Fourmi d'Argentine | Hormiga argentina |
| *Tapinoma melanocephalum* | Ghost ant, Black-headed ant | Schwarzkopfameise, Geisterameise | Fourmi fantôme | Hormiga fantasma, Hormiga bottegaria (Cuba), Albaricoque (Puerto Rico) |
| *Tapinoma sessile* | Odorous house ant | "Wohlriechende Hausameise" | Fourmi odorante | Hormiga doméstica olorosa |
| *Technomyrmex albipes* | White-footed ant | Weissfussameise | — | — |
| *Dorymyrmex* sp. | Pyramid ants | — | — | Hormigas pirámides |

*(continued)*

| Subfamily/Species | Common names | | | |
| --- | --- | --- | --- | --- |
| | English | German | French | Spanish |
| **Myrmicinae** | | | | |
| *Monomorium pharaonis* | Pharaoh ant | Pharao-Ameise | Fourmi pharaon | Hormiga del faraón, Hormiga faraona |
| *Monomorium destructor* | Singapore ant | Singapur-Ameise | — | Hormiga de Singapur |
| *Monomorium floricola* | Bicolored trailing ant | Braunrote Blütenameise | — | — |
| *Monomorium minimum* | Little black ant | Kleine schwarze Ameise | Fourmi petite noire | Hormiga pequeña negra |
| *Pheidole* spp. | Bigheaded ants | Dickkopfameisen | Fourmis à la grosse têtes | Hormigas leonas |
| *Pheidole megacephala* | Coastal brown ant (Australia), Big-headed ant, Madeira house ant | Dickkopfameise | Fourmi à la grosse tête | Hormiga leona |
| *Tetramorium caespitum* | Pavement ant | Rasenameise | Fourmi des gazons | Hormiga del pavimento |
| *Wasmannia auropunctata* | Little fire ant | Rote Feuerameise | — | — |
| *Pogonomyrmex badius* | Florida harvester ant | Ernteameise | — | Hormiga cosechadora |
| *Messor structor* | Harvester ant | Ernteameise | — | — |
| *Solenopsis invicta* | Red imported fire ant | Feuerameise | — | Hormiga de fuego |
| *Solenopsis molesta* | Thief ant, Grease ant | — | Fourmi ravisseuse | Hormiga ladrona |
| *Atta* spp., *Acromyrmex* spp. | Leaf-cutting ants | Blattschneiderameisen | — | Hormigas cortadoras de hojas |
| *Crematogaster* sp. | Acrobat ants | *C. scutellaris* = Rotkopfameise | — | — |
| *Myrmica rubra* | European fire ant | Rote Gartenameise | — | — |

*Notes:* Only economically important genera and species and those with established common names are included. Species that have only an English common name are not listed because these names are mentioned in the text.

164

**APPENDIX 2**

**Urban ant species introduced into Europe**

| Subfamily | Species | Site of introduction (Reference) | Native locality |
|---|---|---|---|
| Formicinae | *Paratrechina longicornis* | U.K. (Cornwell 1978)<br>Germany: Wismar (Steinbrink 2000)<br>Switzerland (Umwelt- und Gesundheitsschutz Zürich 2004)<br>France (Bolton et al. 2006) | Asia or Africa |
| | *Paratrechina vividula* | Germany: Wismar (Steinbrink 2000)<br>Finland (Bolton et al. 2006) | Probably Texas and western Mexico |
| | *Plagiolepis* sp. | Germany: Hamburg, Düsseldorf (Sellenschlo 2002b), Cologne<br>Belgium<br>U.K.: London (Sellenschlo 2002b)<br>Widespread in tropical greenhouses in Europe | Southern Europe and Asia, Africa, and Australia |
| | *Lasius neglectus* | Hungary: Budapest<br>Spain (Rey and Espadaler 2004)<br>Poland, Italy, France (Czechowski et al. 2002) | Asia Minor |
| Dolichoderinae | *Linepithema humile* | Coastal areas of Spain, Portugal, southern France, and northern Italy (Giraud et al. 2002)<br>Mallorca (Sellenschlo 2002b)<br>Greenhouses in Poland (Czechowski et al. 2002)<br>U.K. (Cornwell 1978; C.J. Boase, pers. comm., 2006)<br>Netherlands (M. Brooks, pers. comm., 2008)<br>Germany: Stralsund (Steinbrink 1974), Altenahr (Seifert 1996), Duisburg (R.J. Niessen, pers. comm., 2008) | South America, particularly Argentina, Uruguay, and Brazil |
| | *Tapinoma melanocephalum* | Germany: Cologne, Rostock, Hamburg, Berlin, Neu-Ulm Salzgitter Kempten Halle-Neustadt, and other towns (Scheurer 1984; Sellenschlo 2002b)<br>U.K.: London (C.J. Boase, pers. comm., 2006)<br>Netherlands (M. Brooks, pers. comm., 2007)<br>Poland (Czechowski et al. 2002)<br>Switzerland (Umwelt- und Gesundheitsschutz Zürich 2004) | Africa or the Orient |
| | *Technomyrmex albipes* | Switzerland (Umwelt- und Gesundheitsschutz Zürich 2004)<br>Germany: Bonn (R. Pospischil) | Indo-Pacific region |

*(continued)*

| Subfamily | Species | Site of introduction (Reference) | Native locality |
|---|---|---|---|
| Myrmicinae | *Monomorium pharaonis* | Widely distributed in Europe (only indoors) (Cornwell 1978; Czechowski et al. 2002; Sellenschlo 2002a) | India |
| | *Monomorium floricola* | Germany: Hamburg (Sellenschlo 2002a) | India and Southeast Asia |
| | *Pheidole* spp. | Germany: Wiesbaden (Sellenschlo 2002a) U.K.: London (C.J. Boase, pers. comm., 2006) | |
| | *Pheidole pallidula* | Germany: Berlin (Bauer-Dubau et al. 2001) | Mediterranean region |
| | *Tetramorium insolens* | Mainly in tropical houses in zoological and botanical gardens (Czechowski et al. 2002) | Pacific region |
| | *Tetramorium bicarinatum* | Mainly in tropical houses in zoological and botanical gardens (Czechowski et al. 2002) | Southeast Asia |
| | *Acromyrmex* spp., probably *A. octospinosus* | Germany: Cologne (Pospischil 2007) | Central and South America |
| | *Crematogaster scutellaris* | Germany: Wuppertal (P. Lieving, pers. comm., 2006), northern Germany (Sellenschlo 2002a) Commonly introduced on timber shipments from the South to central Europe (Seifert 1996) | Mediterranean region |
| Ponerinae | *Hypoponera punctatissima* | Poland (tropical greenhouses) (Czechowski et al. 2002), U.K. (Cornwell 1978) | Subtropical origin |

**Urban ant species introduced into the United States**

| Subfamily | Species | Distribution (Reference) | Native locality |
|---|---|---|---|
| Formicinae | *Paratrechina longicornis* | Gulf Coast and Florida (Trager 1984); California (California Academy of Sciences, Antweb) | Asia or Africa |
| | *Paratrechina vividula* | Arizona (Bolton et al. 2006) Southern U.S. from coast to coast (Trager 1984) | Probably native to Texas and western Mexico |
| | *Paratrechina pubens* | Florida: spotty in Miami, also Jacksonville and Sarasota; Texas (Lloyd Davis Jr., pers. comm., 2007); Washington, D.C. (greenhouses) (Trager 1984) | Neotropics |
| | *Paratrechina bourbonica* | Peninsular Florida; Mobile, Alabama (Trager 1984) Hawaii (Bolton et al. 2006) | Tropical tramp species of unknown origin |
| | *Brachymyrmex patagonicus* | Georgia (Ipser et al. 2005); Florida (Deyrup 2003); Florida to Texas (Lloyd Davis Jr., pers. comm., 2007) | Argentina |
| | *Brachymyrmex obscurior* | Georgia (Ipser et al. 2005); Florida (Deyrup 1991) | Neotropics |
| | *Anoplolepis gracilipes* | Hawaii (Reimer et al. 1990) | Africa or tropical Asia |
| | *Plagiolepis alluaudi* | Hawaii, in buildings (Reimer et al. 1990); California: Channel Islands (McGlynn 1999) | Tropical Africa |
| Dolichoderinae | *Linepithema humile* | Throughout southern U.S. but spotty, Maryland, Illinois, Missouri, California, Oregon, Washington, and Hawaii (Vega and Rust 2001) Arizona, Oklahoma (Bolton et al. 2006) | South America, particularly Argentina, Uruguay, and Brazil |
| | *Tapinoma melanocephalum* | Hawaii, Florida, southeastern Texas, California, Pacific Northwest (Hedges 1992, 1997) | Africa or the Orient |
| | *Technomyrmex albipes* | Hawaii, Florida, North Carolina, South Carolina, Georgia, California, and Louisiana (Warner and Scheffrahn 2004a, 2005) | Indo-Pacific region |

*(continued)*

APPENDIX 3—*Continued.*

| Subfamily | Species | Distribution (Reference) | Native locality |
|---|---|---|---|
| Myrmicinae | *Monomorium pharaonis* | Widely distributed in the U.S. (Smith 1965) | India |
| | *Monomorium floricola* | Florida (Vail et al. 1994) and Hawaii (Reimer et al. 1990) | India and Southeast Asia |
| | *Monomorium destructor* | Florida: Key West and Tampa (Deyrup 1991; Vail et al. 1994); Tennessee (Smith 1965) | Africa or India |
| | *Pheidole megacephala* | Southern and central Florida as far north as Inverness (Wilson 2003; Lloyd Davis Jr., pers. comm., 2007); California: Channel Islands (McGlynn 1999) | Africa |
| | *Tetramorium caespitum* | Widely distributed in the U.S. (Bennett et al. 1997; Hedges 1997) | Europe |
| | *Tetramorium insolens* | Sporadic in tropical greenhouses | Pacific region |
| | *Tetramorium bicarinatum* | Long Beach, California, South Carolina, and Florida west to Texas (Martinez 1993) | Southeast Asia |
| | *Tetramorium caldarium* | Florida (McGlynn 1999) | Africa |
| | *Tetramorium simillimum* | Tropical greenhouses (Creighton 1950); yard pests in Florida (Vail et al. 1994) | Africa |
| | *Tetramorium tsushimae* | Missouri and Illinois (Steiner et al. 2006) | East Asia |
| | *Wasmannia auropunctata* | Hawaii, Florida (McGlynn 1999), and California (Ward 2005) | Neotropics |
| | *Solenopsis invicta* | Southern states from Florida to eastern Texas, north into southern Oklahoma, Arkansas, Virginia, and Tennessee; outbreaks in El Paso, Texas, New Mexico, Arizona, California, and Washington (1 record) | Lowland areas of Brazil and Argentina |
| | *Solenopsis richteri* | Northeastern Mississippi, Alabama, and Georgia (Tschinkel 2006; Lloyd Davis Jr., pers. comm., 2007) | Argentina, possibly Brazil |
| | *Myrmica rubra* | Northeastern U.S. and eastern Canada (Groden et al. 2005) | Europe |
| Ponerinae | *Hypoponera punctatissima* | Florida (Vail et al. 1994), Hawaii (McGlynn 1999), Northeast, and Pacific Northwest | Subtropical origin |
| | *Pachycondyla chinensis* | Georgia, Virginia, North Carolina, South Carolina (Nelder et al. 2006) | Southeast Asia |

# References

Abbott, A. 1978. Nutrient dynamics of ants. In M.V. Brian, ed., Production ecology of ants and termites, pp. 233–244. Cambridge: Cambridge University Press.

Adams, E.S., and J.F.A. Traniello. 1981. Chemical interference competition by *Monomorium minimum* (Hymenoptera: Formicidae). Oecologia 51: 265–270.

Akre, R.D. 1992. Thatching ants. Wash. State Univ. Ext. Bull. EB0929.

Akre, R.D., and A.L. Antonelli. 1992. Identification and habits of key ant pests of Washington (workers and winged reproductives). Wash. State Univ. Ext. Bull. EB0671.

Akre, R.D., L.D. Hansen, and E.A. Myhre. 1994a. Colony size and polygyny in carpenter ants (Hymenoptera: Formicidae). J. Kans. Entomol. Soc. 67: 1–9.

Akre, R.D., L.D. Hansen, and E.A. Myhre. 1994b. Do you know where your parents are? Pest Contr. Tech. 22(5): 44, 46, 55, 58, 60, 64.

Akre, R.D., L.D. Hansen, and E.A. Myhre. 1995. Home wreckers! Pest Contr. Tech. 23(1): 54–57, 60, 77.

Alder, P., and J. Silverman. 2005. Effects of interspecific competition between two urban ant species, *Linepithema humile* and *Monormorium minimum,* on toxic bait performance. J. Econ. Entomol. 98: 493–501.

Andersen, A.N. 1991. The ants of Southern Australia—a guide to the Bassian fauna. East Melbourne, Australia: CSIRO Publications.

Anonymous. 1996. Fipronil®. Worldwide Technical Bulletin. Lyon, France: Aventis Crop Science.

Apple, A.G., M.J. Gehret, and M.J. Tanley. 2004. Repellency and toxicity of mint oil granules to red imported fire ants (Hymenoptera: Formicidae). J. Econ. Entomol. 97: 575–580.

Aron, S. 2001. Reproductive strategy: an essential component in the success of incipient colonies of the invasive Argentine ant. Insectes Soc. 48: 25–27.

Baer, H., T.Y. Liu, M.C. Anderson, M. Blum, W.H. Schmid, and F.J. James. 1979. Protein components of fire ant venom (*Solenopsis invicta*). Toxicon 17: 397–405.

Banks, W.A. 1990. Chemical control of the imported fire ants. In R.K. Vander Meer, K. Jaffe, and A. Cedeno, eds., Applied myrmecology: a world perspective, pp. 596–603. Boulder, Colo.: Westview Press.

Banks, W.A., A.S. Las, C.T. Adams, and C.S. Lofgren. 1992. Comparisons of several sulflu-ramid bait formulations for control of the red imported fire ant (Hymenoptera: Formicidae). J. Entomol. Sci. 27: 50–55.

Barbani, L., and R. Fell. 2002. Ants: one hot potato to manage. Pest Contr. 70(3): 20–22.

Barber, E.R. 1920. The Argentine ant as a household pest. USDA Farmer's Bulletin 1101. Washington, D.C.: U.S. Government Printing Office.

Barr, C.L. 2003. Fire ant mound and foraging suppression by indoxacarb bait. J. Agric. Urban Entomol. 20: 143–151.

Bauer-Dubau, K., S. Scheurer, and N. Weidmann. 2001. Erste Erfahrung mit der Bekämpfung der Großkopfameise (*Pheidole pallidula* Nylander) im Land Berlin. Pest Contr. News 28, 30.

Beatson, S.H. 1972. Pharaoh's ants as pathogen vectors in hospitals. The Lancet, Feb. 19, 425–427.

Beatson, S.H. 1973. Pharaoh's ants entering giving-sets. The Lancet, March 17, 606.

Bennett, G.W., J.M. Owens, and R.M. Corrigan. 1997. Truman's scientific guide to pest control operations, 5th ed. Cleveland, Ohio: Advanstar Communications.

Benson, E.P., P.A. Zungoli, and M.B. Riley. 2003. Effects of contaminants on bait acceptance by *Solenopsis invicta* (Hymenoptera: Formicidae). J. Econ. Entomol. 96: 94–97.

Bernstein, R.A. 1979. Schedules of foraging activity in species of ants. J. Anim. Ecol. 48: 921–930.

Bextine, B.R., and H.G. Thorvilson. 2002. Field application of bait-formulated *Beauveria bassiana* alginate pellets for biological control of red imported fire ant (Hymenoptera: Formicidae). Environ. Entomol. 31: 746–752.

Bhatkar, A. 1990. Personal communication with T. Granovsky cited in Mallis Handbook of Pest Control, 7th ed., p. 448. Cleveland, Ohio: Franzak and Foster.

Birkemoe, T. 2002. Structural infestations of ants (Hymenoptera, Formicidae) in southern Norway. Norw. J. Entomol. 49: 139–142.

Blake, C.H. 1940. Notes on economic ants. Pests 8(11): 16–18. Notes on economic ants. Part II. Pests 8(12): 8–10.

Blum, M.S., and H.R. Hermann. 1978. Venoms and venom apparatuses of the Formicidae: Myrmeciinae, Ponerinae, Dorylinae, Pseudomyrmecinae, Myricinae, and Formicinae. In S. Bettini, ed., Arthropod venoms, pp. 801–869. Berlin: Springer Verlag.

Blum, M.S., and E.O. Wilson. 1964. The anatomical source of trail substances in formicine ants. Psyche 71: 28–31.

Bolton, B. 1994. Identification guide to the ant genera of the world. Cambridge, Mass.: Harvard University Press.

Bolton, B. 2003. Synopsis and classification of Formicidae. Mem. Am. Entomol. Inst. 71: 1–370.

Bolton, B., G. Alpert, P.S. Ward, and P. Naskrecki. 2006. Bolton's catalogue of ants of the world 1758–2005 [CD]. Cambridge, Mass.: Harvard University Press.

Bolton, B., and C.A. Collingwood. 1975. Hymenoptera: Formicidae. Handbooks for the identification of British Insects, vol. 6, pt. 3c. London: Royal Entomological Society.

Borgesen, L.W. 2000. Nutritional function of replete workers in the pharaoh's ant, *Monomorium pharaonis* (L.). Insectes Soc. 47: 141–146.

Brett, C.H. 1950. The Texas harvester ant. Okla. Agric. Exp. Sta. Bull. B-353.

Brian, M.V., J. Hibble, and D.J. Stradling. 1965. Ant pattern and density in a southern English heath. J. Anim. Ecol. 34: 545–555.

Brown, W.L. Jr. 1957. Is the ant genus *Tetramorium* native in North America? Breviora 72: 1–8.

Bruder, K.W., and A.P. Gupta. 1972. Biology of the pavement ant, *Tetramorium caespitum* (Hymenoptera: Formicidae). Ann. Entomol. Soc. Am. 65: 358–367.

Buczkowski, G., M.E. Scharf, C.R. Ratliff, and G.W. Bennett. 2005. Efficacy of simulated barrier treatments against laboratory colonies of pharaoh ant. J. Econ. Entomol. 98: 485–492.

Buczkowski, G., E.L. Vargo, and J. Silverman. 2004. The diminutive supercolony: the Argentine ants of the southeastern United States. Mol. Ecol. 13: 2235–2242.

Buren, W.F. 1958. A review of the species of *Crematogaster,* sensu stricto, in North America (Hymenoptera: Formicidae). Part 1. N.Y. Entomol. Soc. 66: 119–134.

Buschinger, A. 2004. International pet ant trade increasing risk and danger in Europe—(Hymenoptera, Formicidae). Aliens 19 and 20: 24–26.

Butovitsch, V. 1976. Über Vorkommen und Schadwirkung der Rossameisen *Camponotus herculeanus* und *C. ligniperda* in Gebäuden in Schweden. Mater. Organismen 11: 161–170.

California Academy of Sciences. Antweb, http://www.antweb.org (accessed 7 February 2008).

Cameron, R.S. 1990. Potential baits for control of the Texas leaf-cutting ant, *Atta texana* (Hymenoptera: Formicidae). In R.K. Vander Meer, K. Jaffe, and A. Cedeno, eds., Applied myrmecology: a world perspective, pp. 628–637. Boulder, Colo.: Westview Press.

Cannon, C.A., and R.D. Fell. 1992. Cold hardiness of the overwintering black carpenter ant. Physiol. Entomol. 17: 121–126.

Cannon, C.A., and R.D. Fell. 2002. Patterns of macronutrient collection in the black carpenter ant, *Camponotus pennsylvanicus* (DeGeer) (Hymenoptera: Formicidae). Environ. Entomol. 31: 977–981.

Caplan, E.L., J.L. Ford, P.F. Young, and D.R. Ownby. 2003. Fire ants represent an important risk for anaphylaxis among residents of an endemic region. J. Allergy Clin. Immunol. 111: 1274–1277.

Causton, C.E., C.R. Sevilla, and S.D. Porter. 2005. Eradication of the little fire ant, *Wasmannia auropunctata* (Hymenoptera: Formicidae), from Marchena Island, Galápagos: on the edge of success? Fla. Entomol. 88: 159–168.

Cherrett, J.M. 1986. The economic importance and control of leaf-cutting ants. In S.B. Vinson, ed., Economic impact and control of social insects, pp. 165–192. New York: Praeger.

Cho, Y.S., Y. Lee, C. Lee, B. Yoo, H. Park, and H. Moon. 2002. Prevalence of *Pachycondyla chinensis* venom allergy in an ant-infested area in Korea. J. Allergy Clin. Immunol. 110: 54–57.

Cohen, S.G., and M. Zelaya-Quesada. 2002. Portier, Richet, and the discovery of anaphylaxis: a centennial. J. Allergy Clin. Immunol. 110: 331–336.

Collingwood, C.A. 1979. The Formicidae (Hymenoptera) of Fennoscandia and Denmark. Fauna Entomol. Scand. 8.

Collins, H.L., and A.-M.A. Callcott. 1995. Effectiveness of spot insecticide treatments for red imported fire ant (Hymenoptera: Formicidae) control. J. Entomol. Sci. 30: 489–496.

Conconi, J.R., M.J. Flores, R.A. Perez, G.L. Cuevas, C.S. Sandoval, C.E. Garduno, I. Portillo, T. Delage-Darchen, and B. Delage-Darchen. 1987. Colony structure of *Liometopum apiculatum* M. and *Liometopum occidentale* var. *luctuosum* W. In J. Eder and H. Rembold, eds., Chemistry and biology of social insects, pp. 671–672. Munich: Verlag J. Peperny.

Cook, T.W. 1953. The ants of California. Palo Alto, Calif.: Pacific Books.

Coovert, G. A. 2005. The ants of Ohio. Ohio Biol. Survey 15(2). Columbus, Ohio: Ohio Biological Survey.

Cornwell, P. 1978. The incidence of pest ants in Britain. Int. Pest Contr. 5: 10–14.

Costa, H.S., L. Greenberg, J. Klotz, and M.K. Rust. 2001. Monitoring the effects of granular insecticides for Argentine ant control in nursery settings. J. Agric. Urban Entomol. 18: 13–22.

Creighton, W.S. 1950. The ants of North America. Bull. Mus. Comp. Zool. Harv. Coll. 104.

Creighton, W.S. 1953. New data on the habits of the ants of the genus *Veromessor.* Am. Mus. Novit. 1612.

Curl, G. 2005. A strategic analysis of the U.S. structural pest control industry: the 2005 season. Gary Curl Specialty Products Consultants, LLC, Mendham, N.J.

Czechowski, W., A. Radchenko, and W. Czechowska. 2002. The ants (Hymenoptera, Formicidae) of Poland. Warsaw: Museum and Institute of Zoology PAS.

Dash, S.T., L.M. Hooper-Bui, and M.A. Seymour. 2005. The pest ants of Louisiana: a guide to their identification, biology and control. Baton Rouge: Louisiana State University Agricultural Center.

David, C.T., and D.L. Wood. 1980. Orientation to trails by a carpenter ant, *Camponotus modoc* (Hymenoptera: Formicidae), in a giant sequoia forest. Can. Entomol. 112: 993–1000.

Davidson, D.W., J.H. Brown, and R.S. Inouye. 1980. Competition and the structure of granivore communities. BioScience 30: 233–238.

Davidson, D.W., S.C. Cook, and R.R. Snelling. 2004. Liquid-feeding performances of ants (Formicidae): ecological and evolutionary implications. Oecologia 139: 255–266.

Deneubourg, J.L., S. Aaron, S. Goss, and J.M. Pasteels. 1990. The self-organizing exploratory pattern of the Argentine ant. J. Insect Behav. 3: 159–168.

deShazo R.D., B.T. Butcher, and W.A. Banks. 1990. Reactions to the stings of the imported fire ant. New Eng. J. Med. 323: 462–466.

Deyrup, M. 1991. Exotic ants of the Florida Keys (Hymenoptera: Formicidae). Proceedings of the Fourth Symposium on the Natural History of the Bahamas, June 7–11, Bahamian Field Station, San Salvador, Bahamas.

Deyrup, M. 2003. An updated list of Florida ants (Hymenoptera: Formicidae). Fla. Entomol. 86(1): 43–48.

Drees, B.M., and R.E. Gold. 2003. Development of integrated pest management programs for the red imported fire ant (Hymenoptera: Formicidae). J. Entomol. Sci. 38: 170–180.

Ebeling, W. 1975. Urban Entomology. Davis: University of California Division of Agricultural Sciences.

Edwards, J.P. 1986. The biology, economic importance, and control of the pharaoh's ant, *Monomorium pharaonis* (L.). In S.B. Vinson, ed., Economic impact and control of social insects, pp. 257–271. New York: Praeger.

Edwards, J.P., and L. Abraham. 1990. Changes in food selection by workers of the pharaoh's ant, *Monomorium pharaonis*. Med. Vet. Entomol. 4: 205–211.

Eisner, T., M. Eisner, and M. Siegler. 2005. Secret weapons. Cambridge, Mass.: Belknap Press of Harvard University Press.

Eisner, T., and G.M. Happ. 1962. The infrabuccal pocket of a formicine ant: a social filtration device. Psyche 69: 107–116.

Eow, A.G., A.S. Chong, and C.Y. Lee. 2004. Colonial growth dynamics of tropical urban pest ants, *Monomorium pharaonis, M. floricola* and *M. destructor* (Hymenoptera: Formicidae). Sociobiology 44: 365–377.

Eow, A.G., A.S. Chong, and C.Y. Lee. 2005. Effects of nutritional starvation and satiation on feeding responses of tropical pest ants, *Monomorium* spp. (Hymenoptera: Formicidae). Sociobiology 45: 15–29.

Eow, A.G.-H., and C.Y. Lee. 2007. Comparative nutritional preferences of tropical pest ants, *Monomorium pharaonis, Monomorium floricola* and *Monomorium destructor* (Hymenoptera: Formicidae). Sociobiology 49: 165–186.

Fernald, H.J. 1947. The little fire ant as a house pest. J. Econ. Entomol. 40: 428.

Fernandes, N., and M.K. Rust. 2003. Site fidelity in foraging Argentine ants (Hymenoptera: Formicidae). Sociobiology 41: 625–632.

Fernández-Meléndez, S., A. Miranda, J.J. García-González, D. Barber, and M. Lombardero. 2007. Anaphylaxis caused by imported red fire ant stings in Málaga, Spain. J. Investig. Allergol. Clin. Immunol. 17(1): 48–49.

Field, H.C., W.E. Evans, R. Hartley, L.D. Hansen, and J.H. Klotz. 2007. A survey of structural ant pests in the southwestern U.S.A. (Hymenoptera: Formicidae). Sociobiology 49: 1–14.

Fisher, B.L., and S.P. Cover. 2007. Ants of North America. Berkeley: University of California Press.

Forschler, B.T. 1997. A prescription for ant control success. Pest Contr. 65(6): 34–38.

Forschler, B.T., and G.M. Evans. 1994. Perimeter treatment strategy using containerized baits to manage Argentine ants, *Linepithema humile* (Mayr) (Hymenoptera: Formicidae). J. Entomol. Sci. 29: 265–267.

Fowler, H.G. 1986. Biology, economics and control of carpenter ants. In S.B. Vinson, ed., Economic impact and control of social insects, pp. 272–289. New York: Praeger.

Fowler, H.G., and R.B. Roberts. 1980. Foraging behavior of the carpenter ant, *Camponotus pennsylvanicus* (Hymenoptera: Formicidae), in New Jersey. J. Kans. Entomol. Soc. 53: 295–304.

Frazier, C.A., and F.K. Brown. 1980. Insects and allergy and what to do about them. Norman: University of Oklahoma Press.

Fukuzawa, M., F. Arakura, Y. Yamazaki, H. Uhara, and T. Saida. 2002. Urticaria and anaphylaxis due to sting by an ant (*Brachyponera chinensis*). Acta. Derm. Venereol. 82: 59.

Furman, B.D., and R.E. Gold. 2006a. Determination of the most effective chemical form and concentration of indoxacarb, as well as the most appropriate grit size, for use in Advion™ Sociobiology 48: 309–334.

Furman, B.D., and R.E. Gold. 2006b. The effectiveness of label-rate broadcast treatment with Advion™ at controlling multiple ant species (Hymenoptera: Formicidae). Sociobiology 48: 559–570.

Gibbons, L., and D. Simberloff. 2005. Interaction of hybrid imported fire ant (*Solenopsis invicta × S. richteri*) with native ants at baits in southeastern Tennessee. Southeast. Nat. 2: 303–320.

Gibson, A. 1916. The control of ants in dwellings—a new remedy. Can. Entomol. 48: 365–366.

Giraud, T., J.S. Pedersen, and L. Keller. 2002. Evolution of supercolonies: the Argentine ants of southern Europe. Proc. Natl. Acad. Sci. USA 99: 6075–6079.

Glancey, B.M., R.K. Vander Meer, A. Glover, C.S. Lofgren, and S.B. Vinson. 1981. Filtration of microparticles from liquids ingested by the red imported fire ant *Solenopsis invicta* Buren. Insectes Soc. 28: 395–401.

Gouge, D.H. 2005. Applications for social insect control. In P.S. Grewal, R.-U. Ehlers, and D.I. Shapiro-Ilan, eds., Nematodes as biocontrol agents, pp. 317–329. Wallingford, U.K.: CAB International.

Granovsky, T.A. 1990. Chapter 12: Ants. In Mallis Handbook of Pest Control, 7th ed., pp. 414–479. Cleveland, Ohio: Franzak and Foster.

Greenberg, L., and J.H. Klotz. 2000. Argentine ant (Hymenoptera: Formicidae) trail pheromone enhances consumption of liquid sucrose solution. J. Econ. Entomol. 93: 119–122.

Greenberg, L., D. Reierson, and M.K. Rust. 2003. Fipronil trials in California against the red imported fire ant, *Solenopsis invicta* Buren, using sugar water consumption and mound counts as measures of ant abundance. J. Agric. Urban Entomol. 20: 221–233.

Greene, A. 2005. Surviving the sting. Pest Contr. Tech. 33(4): 70–72, 75–76.

Gregg, R.E. 1963. The ants of Colorado. Boulder: University of Colorado Press.

Groden, E., F. Drummond, and L.B. Stack. 2004. European fire ant: a new invasive insect in Maine. Univ. Maine Coop. Ext. Bull. 2550.

Groden, E., F.A. Drummond, J. Garnas, and A. Franceour. 2005. Distribution of an invasive ant, *Myrmica rubra* (Hymenoptera: Formicidae), in Maine. J. Econ. Entomol. 98: 1774–1784.

Groden, E., and L.B. Stack. 2008. Managing the invasive European fire ant, *Myrmica rubra*. Orono: Maine Agricultural Center, http://www.mac.umaine.edu/index.php?tab=3&pg= PROJECTS&subaction=getReport&macno=38 (accessed 28 February 2008).

Grosman, D.M., W.W. Upton, F.A. McCook, and R.F. Billings. 2002. Attractiveness and efficacy of fipronil and sulfluramid baits for control of the Texas leafcutting ant. Southwest. Entomol. 27: 251–256.

Gulmahamad, H. 1995. The genus *Liometopum* Mayr (Hymenoptera: Formicidae) in California, with notes on nest architecture and structural importance. Pan-Pac. Entomol. 71(2): 82–86.

Gulmahamad, H. 1996. What have we learned? Pest Contr. 64 (6): 58–59, 62, 64.

Gulmahamad, H. 1997. Argentine ants find southern California hard to resist. Pest Contr. 65(6):72, 73, 76.

Gurney, A.B. 1975. Some stinging ants. Insect World Dig. 2: 19–25.

Haack, K.D., S.B. Vinson, and J.K. Olson. 1995. Food distribution and storage in colonies of *Monomonum pharaonis* (L.) (Hymenoptera: Formicidae). J. Entomol. Sci. 30: 71–81.

Habes, D., S. Morakchi, N. Aribi, J.P. Farine, and N. Soltani. 2006. Boric acid toxicity to the German cockroach, *Blattella germanica:* alterations in midgut structure, and acetylcholinesterase and glutathione *S*-transferase activity. Pest. Biochem. Physiol. 84: 17–24.

Hagen, K.S., R.H. Dadd, and J. Reese. 1984. The food of insects. In C.B. Huffaker and R.L. Rabb, eds., Ecological entomology, pp. 79–112. New York: Wiley.

Haines, I.H., and J.B. Haines. 1978. Colony structure, seasonality and food requirements of the crazy ant, *Anoplolepis longipes* (Jerd.), in the Seychelles. Ecol. Entomol. 3: 109–118.

Haines, I.H., J.B. Haines, and J.M. Cherrett. 1994. The impact and control of the crazy ant, *Anoplolepis longipes* (Jerd.), in the Seychelles. In D.F. Williams, ed., Exotic ants: biology, impact, and control of introduced species, pp. 206–218. Boulder, Colo.: Westview Press.

Hansen, L.D. 1984. A PCO's guide to carpenter ant control. Pest Contr. Tech. 12(4): 56–58.

Hansen, L.D. 1989. Control approach for carpenter ants. Entomology newsletter for county agents. Washington State University Cooperative Extension, January (1): 10.

Hansen, L.D. 2000. Successful bait development is more than a matter of taste. Pest Contr. 68(5): 52, 54, 58.

Hansen, L.D. 2001. Understanding carpenter ant biology and behavior are keys to controlling these widespread terrors. Pest Contr. Tech., Service Technician, May, 25–26.

Hansen, L.D. 2002. Carpenter ant update. Pest Contr. Tech. 30(4): 56, 58, 60, 62, 80.

Hansen, L.D., and R.D. Akre. 1985. Biology of carpenter ants in Washington State (Hymenoptera: Formicidae: *Camponotus*). Melanderia 43: 1–62.

Hansen, L.D., and R.D. Akre. 1990. Biology of carpenter ants. In R.K. Vander Meer, K. Jaffe, and A. Cedeno, eds., Applied myrmecology: a world perspective, pp. 274–280. Boulder, Colo.: Westview Press.

Hansen, L., and J. Klotz. 1999. The name game can wreak havoc on ant control methods. Pest Contr. 67(6): 66, 68, 70.

Hansen, L., and J. Klotz. 2005. Carpenter ants of the United States and Canada. Ithaca: Cornell University Press.

Harada, A.Y. 1990. Ant pests of the Tapinomini tribe. In R.K. Vander Meer, K. Jaffe, and A. Cedeno, eds., Applied myrmecology: a world perspective, pp. 298–309. Boulder, Colo.: Westview Press.

Headley, A.E. 1949. A population study of the ant, *Aphaenogaster fulva* subsp. *aquia* Buckley. Ann. Entomol. Soc. Am. 42: 265–272.

Hedges, S.A. 1992. Field guide for the management of structure-infesting ants. Cleveland, Ohio: Franzak and Foster.

Hedges, S.A. 1997. Chapter 12: Ants. In Handbook of Pest Control, 8th ed., pp. 503–589. Cleveland, Ohio: Mallis Handbook and Technical Training Co.

Hedges, S.A. 1998. Field guide for the management of structure-infesting ants. 2nd ed. Cleveland, Ohio: Franzak and Foster.

Hedges, S.A. 2002. 2002 pest wrap-up. Pest Contr. Tech. 30(12): 56, 58–60, 86.

Hedlund, K. 2002. The Ants: genus *Myrmica* (Myrmicinae). Chapel Hill: University of North Carolina, http://www.cs.unc.edu/~hedlund/dev/ants/catalog/ (accessed 26 August 2004).

Hedlund, K. 2003. Online catalog of the North American ants. Chapel Hill: University of North Carolina, http://www.cs.unc.edu/~hedlund/ants/ (accessed 20 October 2007).

Heraty, J. 1994. Biology and importance of two eucharitid parsites of *Wasmannia* and *Solenopsis*. In D.F. Williams, ed., Exotic ants: biology, impact, and control of introduced species, pp. 104–120. Boulder, Colo.: Westview Press.

Herms, W.B. 1939. Medical entomology. New York: Macmillan.

Higgins, W., D. Bell, C. Silcox, and G. Holbrook. 2002. Liquid bait formulations for controlling the odorous house ant (Hymenoptera: Formicidae). In S.C. Jones, J. Zhai, and W.H. Robinson, eds., Proceedings of the Fourth International Conference on Urban Pests, pp. 129–134. Blacksburg, Va.: Pocahantas Press.

Hoffman, D.R. 1997. Reactions to less common species of fire ants. J. Allergy Clin. Immunol. 100: 679–683.

Hölldobler, B. 1982. Interference strategy of *Iridomyrmex pruinosum* (Hymenoptera: Formicidae) during foraging. Oecologia 52: 208–213.

Hölldobler, B., and E.O. Wilson. 1990. The ants. Cambridge, Mass.: Harvard University Press.

Hölldobler, B., and E.O. Wilson. 1994. Journey to the ants. Cambridge, Mass.: Harvard University Press.

Hollinghaus, J.G. 1987. Inhibition of mitochondrial electron transport by hydramethylnon: a new amidinohydrazone insecticide. Pest. Biochem. Physiol. 27: 61–70.

Holway, D.A., and T.J. Case. 2000. Mechanisms of dispersed central-place foraging in polydomous colonies of the Argentine ant. Anim. Behav. 59: 433–441.

Holway, D.A., L. Lach, A.V. Suarez, N.D. Tsutsui, and T.J. Case. 2002. The causes and consequences of ant invasions. Annu. Rev. Entomol. 33: 181–233.

Hooper, L.M., and M.K. Rust. 1997. Food preference and patterns of foraging activity of the southern fire ant (Hymenoptera: Formicidae). Ann. Entomol. Soc. Am. 90: 246–253.

Hooper, L.M., M.K. Rust, and D.A. Reierson. 1998. Using bait to suppress the southern fire ant on an ecologically sensitive site (Hymenoptera: Formicidae). Sociobiology 31: 283–289.

Hooper-Bui, L.M., A.G. Appel, and M.K. Rust. 2002. Preference of food particle size among several urban ant species. J. Econ. Entomol. 95: 1222–1228.

Horton, J.R. 1918. The Argentine ant in relation to citrus groves. USDA Agric. Bull. 647. Washington, D.C.: U.S. Department of Agriculture.

Howard, D.F., and W.S. Tschinkel. 1981. The flow of food in colonies of the fire ant, *Solenopsis invicta;* a multifactorial study. Physiol. Entomol. 6: 297–306.

Human, K.G., and D.M. Gordon. 1996. Exploitation and interference competition between the invasive Argentine ant, *Linepithema humile,* and native ant species. Oceologia 105: 405–412.

Ipser, R.M., M.A. Brinkman, and W.A. Gardner. 2005. First report of *Brachymyrmex obscurior* Forel (Hymenopera: Formicidae) from Georgia. J. Entomol. Sci. 40: 250–251.

Jackson, D.E., M. Holcombe, and F.L.W. Ratnieks. 2004. Trail geometry gives polarity to ant foraging networks. Nature 432: 907–909.

Jacobson, A.L., D.C. Thompson, L. Murray, and S.F. Hanson. 2006. Establishing guidelines to improve identification of fire ants *Solenopsis xyloni* and *Solenopsis invicta*. J. Econ. Entomol. 99: 313–322.

Jander, R., and U. Jander. 1998. The light and magnetic compass of the weaver ant, *Oecophylla smaragdina* (Hymenoptera: Formicidae). Ethology 104: 743–758.

Japanese Ant Image Database. 2003. Ant Database Group, http://ant.edb.miyakyo-u.ac.jp/E/ (accessed 28 January 2008).

Jetter, K.M., J. Hamilton, and J.H. Klotz. 2002. Red imported fire ants threaten agriculture, wildlife and homes. Calif. Agric. 56 (1): 26–34.

Josens, R.B., W.M. Farnia, and F. Roces. 1998. Nectar feeding by the ant *Camponotus mus:* intake rate and crop filling as a function of sucrose concentration. J. Insect Physiol. 44: 579–585.

Jusino-Atresmo, R., and S.A. Phillips Jr. 1994. Impact of red imported fire ants on the ant fauna of central Texas. In D.F. Williams, ed., Exotic ants: biology, impact, and control of introduced species, pp. 259–268. Boulder, Colo.: Westview Press.

Kaufmann, B., J.J. Boomsma, L. Passera, and K.N. Petersen. 1992. Relatedness and inbreeding in a French population of the unicolonial ant *Iridomyrmex humilis* (Mayr). Insectes Soc. 39: 195–213.

Kay, A. 2004. The relative availabilities of complimentary resources affect the feeding preferences of ant colonies. Behav. Ecol. 15: 63–70.

Keck, M.E., R.E. Gold, and S.B. Vinson. 2005. Invasive interactions of *Monomorium minimum* (Hymenoptera: Formicidae) and *Solenopsis invicta* (Hymenoptera: Formicidae) infected with *Thelohania solenopsae* (Microsporida: Thelohaniidae) in the laboratory. Sociobiology 46: 73–86.

Keller, L., L. Passera, and J.P. Suzzoni. 1989. Queen execution in the Argentine ant *Iridomyrmex humilis* (Mayr). Physiol. Entomol. 14: 157–163.

Kemp, S.F., R.D. de Shazo, J.E. Moffitt, D.F. Williams, and W.A. Buhner. 2000. Expanding habitat of the imported fire ant (*Solenopsis invicta*): a public health concern. J. Allergy Clin. Immunol. 105: 683–691.

Kim, S., and C. Hong. 1992. A case of anaphylaxis by ant (*Ectomomyrmex* spp.) venom and measurements of specific IgE and IgG subclasses. Yonsei Med. J. 33: 281–287.

Kim, S.S., H.S. Park, H.Y. Kim, S.K. Lee, and D.H. Nahm. 2001. Anaphylaxis caused by the new ant, *Pachycondyla chinensis:* demonstration of specific IgE and IgE-binding components. J. Allergy Clin. Immunol. 107: 1095–1099.

Klein, R.W. 1987. Colony structures of three species of *Pseudomyrmex* (Hymenoptera: Formicidae: Pseudomyrmecinae) in Florida. In J. Eder and H. Rembold, eds., Chemistry and biology of social insects, pp. 107–108. Munich: Verlag J. Peperny.

Klotz, J.H., C. Amrhein, S. McDaniel, M.K. Rust, and D.A. Reierson. 2002. Assimilation and toxicity of boron in the Argentine ant (Hymenoptera: Formicidae). J. Entomol. Sci. 37:193–199.

Klotz, J.H., R.D. deShazo, J.L. Pinnas, A.M. Frishman, J.O. Schmidt, D.R. Suiter, G.W. Price, and S.A. Klotz. 2005. Adverse reactions to ants other than imported fire ants. Ann. Allergy Asthma Immunol. 95: 418–425.

Klotz, J.H., H.C. Field, S.A. Klotz, and J.L. Pinnas. 2006. Ants and public health. Pest Contr. Tech. 34(3): 40, 42, 44, 46, 48, 50–52.

Klotz, J.H., L. Greenberg, H.H. Shorey, and D.F. Williams. 1997a. Alternative strategies for ants around homes. J. Agric. Entomol. 14: 249–257.

Klotz, J.H., J.R. Mangold, K.M. Vail, L.R. Davis Jr., and R.S. Patterson. 1995. A survey of the urban ant pests (Hymenoptera: Formicidae) of peninsular Florida. Fla. Entomol. 78: 109–118.

Klotz, J., B. Reid, and J. Hamilton. 2000. Locomotory efficiency in ants using structural guidelines (Hymenoptera: Formicidae). Sociobiology 35: 79–88.

Klotz, J.H., M.K. Rust, H.S. Costa, D.A. Reierson, and K. Kido. 2002. Strategies for controlling Argentine ants (Hymeoptera: Formicidae) with sprays and baits. J. Agric. Urban Entomol. 19: 85–94.

Klotz, J.H., M.K. Rust, L. Greenberg, H.C. Field, and K. Kupfer. 2007. An evaluation of several urban pest management strategies to control Argentine ants (Hymenoptera: Formicidae). Sociobiology 50: 1–8.

Klotz, J.H., J.O. Schmidt, J.L. Pinnas, and S.A. Klotz. 2005. Consequences of harvester ant incursion into urbanized areas: a case history of sting anaphylaxis (Hymenoptera: Formicidae). Sociobiology 45: 543–551.

Klotz, J.H., K.M. Vail, and D.F. Williams. 1997b. Toxicity of a boric acid—sucrose water bait to *Solenonsis invicta* (Hymenoptera: Formicidae). J. Econ. Entomol. 90: 488–491.

Klotz, S., J. Schmidt, R. Kohlmeier, D. Suiter, J. Pinnas, and J. Klotz. 2004. Stinging ants: case histories of three native North American species. In T. Sutphin, D. Miller, and R. Kopanic, eds., Proceedings of the 2004 National Conference on Urban Entomology, Phoenix, Arizona, 20–22 May 2004, pp. 108–109.

Knight, R.L., and M.K. Rust. 1990a. The urban ants of California with distribution notes of imported species. Southwest. Entomol. 15: 167–178.

Knight, R.L., and M.K. Rust. 1990b. Repellency and efficacy of various insecticides against foraging workers in laboratory colonies of the Argentine ant, *Iridomyrmex humilis* (Mayr) (Hymenoptera: Formicideae). J. Econ. Entomol. 83: 1402–1408.

Knight, R.L., and M.K. Rust. 1991. Efficacy of formulated baits for control of Argentine ant (Hymenoptera: Formicidae). J. Econ. Entomol. 84: 510–514.

Krieger, M.J.B., and L. Keller. 2000. Mating frequency and genetic structure of the Argentine ant, *Linepithema humile*. Mol. Ecol. 9: 119–126.

Kuby, J. 1991. Immunology, 5th ed. New York: W.H. Freeman.

Leath, T.M., T.J. Grier, R.S. Jacobson, and M.E. Fontana-Penn. 2006. Anaphylaxis to *Pachycondyla chinensis* [abstract]. J. Allergy Clin. Immunol. 117: S129.

Lennartz, F.E. 1973. Modes of dispersal of *Solenopsis invicta* from Brazil into the continental United States—a study in spatial diffusion. MS thesis, University of Florida, Gainesville.

Lim, S.P., A.S.C. Chong, and C.Y. Lee. 2003. Nestmate recognition and intercolonial aggression in the crazy ant, *Paratrechina longicornis* (Hymenoptera: Formicidae). Sociobiology 41: 295–305.

Lofgren, C.S. 1986. The economic importance and control of imported fire ants in the United States. In S.B. Vinson, ed., Economic impact and control of social insects, pp. 227–256. New York: Praeger.

Lofgren, C.S., W.A. Banks, and B.M. Glancey. 1975. Biology and control of imported fire ants. Annu. Rev. Entomol. 20: 1–30.

Lofgren, C.S., and D.F. Williams. 1982. Avermectin $B_1a$: highly potent inhibitor of reproduction by queens of the red imported fire ant (Hymenoptera: Formicidae). J. Econ. Entomol. 75: 798–803.

Loftin, I.C., J. Hopkins, J. Gavin, and D. Shanklin. 2003. Evaluation of broadcast applications of various contact insecticides against red imported fire ants, *Solenopsis invicta* Buren. J. Agric. Urban Entomol. 20: 151–156.

Lopez, R., D.W. Held, and D.A. Potter. 2000. Management of a mound-building ant, *Lasius neoniger* Emery, on golf putting greens and tees using delayed-action baits or fipronil. Crop Sci. 40: 511–517.

MacKay, W.P., L. Greenberg, and S.B. Vinson. 1994. A comparison of bait recruitment in monogynous and polygynous forms of the red imported fire ant, *Solenopsis invicta* Buren. J. Kans. Entomol. Soc. 67: 133–136.

Maier, R.M., and D.A. Potter. 2005. Seasonal mounding, colony development, and control of nuptial queens of the ant *Lasius neoniger*. Applied Turfgrass Sci., http://www.plant managementnetwork.org/ats/, doi:10.1094/ATS-2005-0502-01-RS (accessed 28 January 2008).

Majer, J.D. 1994. Spread of Argentine ants (*Linepithema humile*), with special reference to

Western Australia. In D.F. Williams, ed., Exotic ants: biology, impact, and control of introduced species, pp. 163–173 Boulder, Colo.: Westview Press.

Majeski, J.A., G.G. Durst, and K.T. McKee. 1974. Acute systemic anaphylaxis associated with an ant sting. South. Med. J. 67: 365–366.

Mallis, A. 1969. Handbook of pest control, 5th ed. New York: MacNair-Dorland.

Mangold, J. 1996. Personal communication with Stoy Hedges. Cited in Handbook of Pest Control, 8th ed., p. 545. Cleveland, Ohio: Mallis Handbook and Technical Training Co.

Markin, G.P. 1967. Food distribution within colonies of the Argentine ant, *Iridomyrmex humilis* (Mayr). PhD dissertation, University of California, Riverside.

Markin, G.P. 1968. Nest relationship of the Argentine ant, *Iridomyrmex humilis* (Hymenoptera: Formicidae). J. Kans. Entomol. Soc. 41: 511–516.

Markin, G.P. 1970a. The seasonal life cycle of the Argentine ant, *Iridomyrmex humilis* (Hymenoptera: Formicidae), in southern California. Ann. Entomol. Soc. Am. 63: 1238–1242.

Markin, G.P. 1970b. Foraging behavior of the Argentine ant in a California citrus grove. J. Econ. Entomol. 63: 741–744.

Markin, G.P. 1970c. Food distribution within laboratory colonies of the Argentine ant, *Iridomyrmex humilis* (Mayr). Insectes Soc. 17: 127–158.

Martinez, M.J. 1993. The first field record for the ant *Tetramorium bicarinatum* Nylander (Hymenoptera: Formicidae) in California. Pan-Pac. Entomol. 69: 272–273.

Martinez, M.J. 1995. The first record of mixed nests of *Conomyrma bicolor* (Wheeler) and *Conomyrma insana* (Buckley) (Hymenoptera: Formicidae). Pan-Pac. Entomol. 71: 252.

McDaniel, S.G., and W.L. Sterling. 1979. Predator determination and efficiency on *Heliothis virescens* eggs in cotton using [32]P. Environ. Entomol. 8: 1083–1087.

McDaniel, S.G., and W.L. Sterling. 1982. Predation of *Heliothis virescens* (F.) eggs on cotton in east Texas. Environ. Entomol. 11: 60–66.

McGlynn, T.P. 1999. The worldwide transfer of ants: geographical distribution and ecological invasions. J. Biogeogr. 26: 535–548.

McIver, J.D. 1991. Dispersed central place foraging in Australian meat ants. Insectes Soc. 38: 129–137.

Meier, R.E. 1994. Coexisting patterns and foraging behavior of introduced and native ants (Hymenoptera: Formicidae) in the Galápagos Islands (Ecuador). In D.F. Williams, ed., Exotic ants: biology, impact, and control of introduced species, pp. 44–62. Boulder, Colo.: Westview Press.

Meissner, H.E., and J. Silverman. 2001. Effects of aromatic cedar mulch on the Argentine ant and the odorous house ant (Hymenoptera: Formicidae). J. Econ. Entomol. 94: 1526–1531.

Meissner, H.E., and J. Silverman. 2003. Effect of aromatic cedar mulch on Argentine ant (Hymenoptera: Formicidae) foraging activity and nest establishment. J. Econ. Entomol. 96: 850–855.

Merchant, M., and B.M. Drees. 1992. The two-step method do-it-yourself fire ant control. Tex. Agric. Ext. Serv. Publ. L-5070.

Merck. 1976. The Merck Index, 9th ed. Rahway, N.J.: Merck and Co.

Merickel, F.W., and W.H. Clark. 1994. *Tetramorium caespitum* (Linnaeus) and *Liometopum luctuosum* W.M. Wheeler (Hymenoptera: Formicidae): new state records for Idaho and Oregon, with notes on their natural history. Pan-Pac. Entomol. 70: 148–158.

Michener, C.D. 1942. The history and behavior of a colony of harvesting ants. The Scientific Monthly 55: 248–258.

Moffitt, J.E. 2004. Reactions to insect bites and stings: what about the orphan insects? Ann. Allergy, Asthma, Immunol. 93: 507–509.

Moreau, C.S., C.D. Bell, R. Vila, S.B. Archibald, and N.E. Pierce. 2006. Phylogeny of the ants: diversification in the age of angiosperms. Science 312: 101–104.

Morgan, E.D., B.D. Jackson, and J. Billen. 2005. Chemical secretions of the "crazy ant" *Paratrechina longicornis* (Hymenoptera: Formicidae). Sociobiology 46: 299–304.

Morrison, L.W., and S.D. Porter. 2005. Testing for population-level impacts of introduced *Pseudoacteon tricuspis* flies, phorid parasitoids of *Solenopsis invicta* fire ants. Biol. Contr. 33: 9–19.

Moser, J.C. 1967. Mating activities of *Atta texana* (Hymenoptera: Formicidae). Insectes Soc. 14: 295–312.

Moser, J.C. 2006. Complete excavation and mapping of a Texas leafcutting ant nest. Ann. Entomol. Soc. Am. 99: 891–897.

Nelder, M.P., E.S. Paysen, P.A. Zungoli, and E.P. Benson. 2006. Emergence of the introduced ant *Pachycondyla chinensis* (Formicidae: Ponerinae) as a public health threat in the southeastern United States. J. Med. Entomol. 43: 1094–1098.

Newell, W. 1909. Measures suggested against the Argentine ant as a household pest. J. Econ. Entomol. 2: 324–332.

Newell, W., and T.C. Barber. 1913. The Argentine ant. USDA Bur. Entomol. Bull. 122.

Nickerson, J.C., and K.A. Barbara. 2000. Featured creatures: crazy ant. Gainesville: University of Florida Institute of Food and Agricultural Sciences, http://creatures.ifas.ufl.edu (accessed 20 December 2006).

Nickerson, J.C., and C.L. Bloomcamp. 2003. Featured creatures: ghost ant. Gainesville: University of Florida Institute of Food and Agricultural Sciences, http://creatures.ifas.ufl.edu (accessed 20 December 2006).

Nugent J.S., D.R. More, L.L. Hagan, J.G. Demain, B.A. Whisman, and T.M. Freeman. 2004. Cross-reactivity between allergens in the venom of the common striped scorpion and the imported fire ant. J. Allergy Clin. Immunol. 114: 383–386.

Nuss, A.B., D.R. Suiter, and G.W. Bennett. 2005. Continuous monitoring of the black carpenter ant, *Camponotus pennsylvanicus* (Hymenoptera: Formicidae), trail behavior. Sociobiology 45: 597–618.

O'Brien, K.S., and L.M. Hooper-Bui. 2005. Hunger in red imported fire ants and their behavioral response to two liquid bait products. J. Econ. Entomol. 98: 2153–2159.

Oi, D.H. 2002. Biological and chemical control of ants. Report to the Almond Board of California. Project no. 2001-DO-00.

Oi, D.H. 2006. Effect of mono- and polygyne social forms on transmission and spread of a microspordium in fire ant populations. J. Invert. Pathol. 92: 146–151.

Oi, D.H., J.A. Briano, S.M. Valles, and D.F. Williams. 2005. Transmission of *Vairimorpha invictae* (Microsporidia: Burenellidae) infections between red imported fire ant (Hymenoptera: Formicidae) colonies. J. Invert. Pathol. 88: 108–115.

Oi, D.H., and F.M. Oi. 2006. Speed of efficacy and delayed toxicity characteristics of fast-acting fire ant (Hymenoptera: Formicidae) baits. J. Econ. Entomol. 99: 1739–1748.

Oi, D.H., K.M. Vail, and D.F. Williams. 1996. Field evaluation of perimeter treatments for pharaoh ant (Hymenoptera: Formicidae) control. Fla. Entomol. 79: 252–263.

Oi, D.H., K.M. Vail, and D.F. Williams. 2000. Bait distribution among multiple colonies of pharaoh ants (Hymenoptera: Formicidae). J. Econ. Entomol. 93: 1247–1255.

Oi, D.H., K.M. Vail, D.F. Williams, and D.N. Bieman. 1994. Indoor and outdoor foraging locations of pharaoh ants (Hymenoptera: Formicidae) and control strategies using bait stations. Fla. Entomol. 77: 85–91.

Orr, M.R., S.H. Seike, W.W. Benson, and D.L. Dahlsten. 2001. Host specificity of *Pseudacteon* (Diptera: Phoridae) parasitoids that attack *Linepithema* (Hymenoptera: Formicidae) in South America. Environ. Entomol. 30: 742–747.

Passera, L. 1994. Characteristics of tramp species. In D.F. Williams, ed., Exotic ants: biology, impact, and control of introduced species, pp. 23–43. Boulder, Colo.: Westview Press.

Paul, J., and F. Roces. 2003. Fluid intake rates in ants correlate with their feeding habits. J. Insect Physiol. 49: 347–357.

Paysen, E., P. Zungoli, and E. Benson. 2007. Packing a punch. Pest Contr. Tech. 35(4): 55–56, 58, 60.

Pereira, R.M. 2003. Areawide suppression of fire ant populations in pastures: project update. J. Agric. Urban Entomol. 20: 123–130.

Pereira, R.M., and J.L. Stimac. 1997. Biocontrol options for urban pest ants. J. Agric. Entomol. 14: 231–248.

Petal, J. 1978. The role of ants in ecosystems. In M.V. Brian, ed., Production ecology of ants and termites, pp. 293–325. International Biological Programme 13. London: Cambridge University Press.

Peters, A. 1996. The natural host range of *Steinernema* and *Heterorhabditis* spp. and their impact on insect populations. Biocontr. Sci. Tech. 6: 389–402.

Pinnas, J.L. Allergic reactions to insect stings. 2001. In R.E. Rakel and E.T. Bope, eds., Conns' current therapy, pp. 797–799. Philadelphia: W.B. Saunders.

Pinnas, J.L., R.C. Strunk, T.M. Wang, and H.C. Thompson. 1977. Harvester ant sensitivity: in vitro and in vivo studies using whole body extracts and venom. J. Allergy Clin. Immunol. 59: 10–16.

Porter, S.D., and D.A. Savignano. 1990. Invasion of polygyne fire ants decimates ants and disrupts arthropod community. Ecology 71: 2095–2106.

Porter, S.D., and W.R. Tschinkel. 1987. Foraging in *Solenopsis invicta* (Hymenoptera: Formicidae): effects of weather and season. Environ. Entomol. 16: 802–808.

Pospischil, R. 2005. Ameisen—Lebensweise und Bekämpfung. [Ants—biology and control in food production.] Der Lebensmittelbrief 14, 9–10, 201–206.

Pospischil, R. 2007. Blattschneiderameisen. Der praktische Schädlingsbekämpfer 59(3): 14–15.

Pospischil, R. 2008. Neu: Maxforce Quantum, ein Flüssigköder gegen verschiedene Ameisenarten. Der praktische Schädlingsbekämpfer, 60(2): 20–21.

Potter, M., and A. Hillery. 2003. Nonrepellent strategies for OHA. Pest Contr. 71: 32–34.

Pranschke, A.M., L.M. Hooper-Bui, and B. Moser. 2003. Efficacy of bifenthrin treatment zones against red imported fire ant. J. Econ. Entomol. 96: 98–105.

Pricer, J.L. 1908. The life history of the carpenter ant. Biol. Bull. 14: 177–218.

Reagan, T.E. 1981. Sugarcane borer pest management in Louisiana: leading to a more permanent system. Proceedings of the Second Inter-American Sugarcane Seminar: Insect and Rodent Pests, Florida International University, Miami, Florida, Oct. 1981, pp. 100–110.

Reierson, D.A., M.K. Rust, and J. Hampton-Beesley. 1998. Monitoring with sugar water to determine the efficacy of treatments to control Argentine ants, *Linepithema humile* (Mayr). In Proceedings of the 1998 National Conference on Urban Entomology, pp. 78–82.

Reierson, D.A., M.K. Rust, and J. Klotz. 2001. There's safety in numbers. Pest Contr. 69(3): 50–52.

Reimer, N., J.W. Beardsley, and G. Jahn. 1990. Pest ants in the Hawaiian Islands. In R.K. Vander Meer, K. Jaffe, and A. Cedeno, eds., Applied myrmecology: a world perspective, pp. 40–50. Boulder, Colo.: Westview Press.

Rey, S., and X. Espadaler. 2004. Area-wide management of the invasive garden ant *Lasius neglectus* (Hymenoptera: Formicidae) in northeast Spain. J. Agric. Urban Entomol. 21: 99–112.

Richman, D.L., and L.M. Hooper-Bui. 2003. Mystery solved? Pest Contr. Tech. 31: 84, 86–87.

Riggs, N.L., L. Lennon, C.L. Barr, B.M. Drees, S. Cummings, and C. Lard. 2002. Community-wide red imported fire ant management programs in Texas. Southwest. Entomol., Suppl. 25: 31–41.

Rissing, S.W. 1988. Group foraging dynamics of the desert seed-harvester ant *Veromessor pergandei* (Mayr). In J.C. Trager, ed., Advances in myrmecology, pp. 347–353. New York: E.J. Brill.

Robinson, E.J.H., D.E. Jackson, M. Holcombe, and F.L.W. Ratnieks. 2005. "No entry" signal in ant foraging. Nature 438: 442.

Rust, M.K. 1986. Managing household pests. In G.W. Bennett and J.M. Owens, eds., Advances in urban pest management, pp. 335–368. New York: Van Nostrand Reinhold.

Rust, M.K., K. Haagsma, and D.A. Reierson. 1996. Barrier sprays to control Argentine ants (Hymenoptera: Formicidae). J. Econ. Entomol. 89:134–137.

Rust, M.K., and R.L. Knight. 1990. Controlling Argentine ants in urban situations. In R.K. Vander Meer, K. Jaffe, and A. Cedeno, eds., Applied myrmecology: a world perspective, pp. 663–670. Boulder, Colo.: Westview Press.

Rust, M.K., D.A. Reierson, and J.H. Klotz. 2002. Factors affecting the performance of bait toxicants for Argentine ants (Hymenoptera: Formicidae). In S.C. Jones, J. Zhai, and W.H. Robinson, eds., Proceedings of the Fourth International Conference on Urban Pests, pp. 115–120. Blacksburg, Va.: Pocahontas Press.

Rust, M.K., D.A. Reierson, and J.H. Klotz. 2003. Pest management of Argentine ants (Hymenoptera: Formicidae). J. Entomol. Sci. 38: 159–169.

Rust, M.K., D.A. Reierson, and J.H. Klotz. 2004. Delayed toxicity as a critical factor in the efficacy of aqueous baits for controlling Argentine ants (Hymenoptera: Formicidae). J. Econ. Entomol. 97: 1017–1024.

Rust, M.K., D.A. Reierson, E. Paine, and L.J. Blum. 2000. Seasonal activity and bait preferences of the Argentine ant (Hymenoptera: Formicidae). J. Agric. Urban Entomol. 17: 201–212.

Salgado, V.L., J.L. Sheets, G.B. Watson, and A.L. Schmidt. 1998. Studies on the mode of action of spinosad: the internal effective concentration and the concentration dependence of neural excitation. Pest. Biochem. Physiol. 60: 103–110.

Samter, M. 1969. Anaphylaxis. In M. Samter, ed., Excerpts from classics in allergy, pp. 32–34. Columbus, Ohio: Ross Laboratories.

Sanders, C.J. 1970. Distribution of carpenter ant colonies in the spruce-fir forests of northwestern Ontario. Ecology 51: 865–873.

Scharf, M.E., C.R. Ratliff, and G.W. Bennett. 2004. Impacts of residual insecticide barriers on perimeter-invading ants, with particular reference to the odorous house ant, *Tapinoma sessile*. J. Econ. Entomol. 97: 601–605.

Scheurer, S. 1984. Erstnachweis des Hygieneschädlings *Tapinoma melanocephalum* (Hym., Formicidae) in der DDR. Angew. Parasitol. 25: 96–99.

Schmid-Grendelmeier, P., M. Lundberg, and B. Wüthrich. 1997. Anaphylaxis due to a red harvester ant bite. Allergy 52: 230–231.

Schmidt, J.O. 1982. Biochemistry of insect venoms. Ann. Rev. Entomol. 27: 339–368.

Schmidt, J.O. 1986. Chemistry, pharmacology, and chemical ecology of ant venoms. In T. Piek, ed., Venoms of the Hymenoptera: biochemical, pharmacological and behavioural aspects, pp. 425–508. Orlando, Fla.: Academic Press.

Schmidt, J.O. 2003. Venom. In R.H. Resh and R.T. Cardé, eds., Encyclopedia of insects, pp. 1160–1163. Amsterdam: Academic Press.

Schmidt, J.O., G.C. Meinke, T.M. Chen, and J.L. Pinnas. 1984. Demonstration of cross-allergenicity among harvester ant venoms using RAST and RAST inhibition [abstract]. J. Allergy Clin. Immunol. 73: 158.

Schmidt, P.J., W.C. Sherbrooke, and J.O. Schmidt. 1989. The detoxification of ant (*Pogonomyrmex*) venom by a blood factor in horned lizards (*Phrynosoma*). Copeia 1989: 603–607.

Schneirla, T.C. 1958. The behavior and biology of certain Nearctic army ants, last part of the functional season, southeastern Arizona. Insectes Soc. 5: 215–255.

Schultz, D.R., J.J. Byrnes, and H.E. Brown. 1978. Response of mixed cryoglobulinemia to treatment with ant venom [abstract]. Clin. Res. 26: 58A.

Seifert, B. 1988. A taxonomic revision of the *Myrmica* species of Europe, Asia Minor, and Caucasia (Hymenoptera, Formicidae). Abh. Ber. Naturkundemus, Görlitz 62: 1–75.

Seifert, B. 1992. A taxonomic revision of the Palaearctic members of the ant subgenus *Lasius* s.str. (Hymenoptera: Formicidae). Abh. Ber. Naturkundemus, Görlitz 66: 1–67.

Seifert, B. 1996. Ameisen—beobachten, bestimmen. Augsburg, Germany: Weltbild Verlag.

Seifert, B. 2007. Die Ameisen Mittel- und Nordeuropas. Görlitz/Tauer, Germany: Lutra Verlag.

Sellenschlo, U. 2002a. Eingeschleppte Ameisen, part 2. Der praktische Schädlingsbekämpfer 54(2): 10–11.

Sellenschlo, U. 2002b. Eine Überraschung im Gepäck—Eingeschleppte Ameisenarten und ihre Bestimmung, part 3. Der praktische Schädlingsbekämpfer 54(3): 22–24.

Shattuck, S.O., and N.J. Barnett. 2001. Australian ants online: the guide to the Australian ant fauna. CSIRO Australia, http://www.ento.csiro.au/science/ants (accessed 24 April 2001).

Shetlar, D.J., and V.E. Walter. 1982. Chapter 14: Ants. In Mallis Handbook of Pest Control, 6th ed., pp. 424–487. Cleveland, Ohio: Franzak and Foster.

Silverman, J., and T.H. Roulston. 2001. Acceptance and intake of gel and liquid sucrose compositions by the Argentine ant (Hymenoptera: Formicidae). J. Econ. Entomol. 94: 511–515.

Silverman, J., and T.H. Roulston. 2003. Retrieval of granular bait by the Argentine ant (Hymenoptera: Formicidae): effect of clumped versus scattered dispersion patterns. J. Econ. Entomol. 96: 871–874.

Silverman, J., C.E. Sorenson, and M.G. Waldvogel. 2006. Trap-mulching Argentine ants. J. Econ. Entomol. 99: 1757–1760.

Smith, L.M., A.G. Appel, and G.J. Keever. 1996. Cockroach control methods can cause other pest problems. Ala. Agric. Exp. Sta., Highlights Agric. Res. 43: 5–6.

Smith, M.R. 1928. The biology of *Tapinoma sessile* Say, an important house-infesting ant. Ann. Entomol. Soc. Am. 21: 307–329.

Smith, M.R. 1965. House-infesting ants of the eastern United States: their recognition, biology, and economic importance. Agric. Tech. Bull. 1326. Washington, D.C.: U.S. Department of Agriculture, Agricultural Research Service.

Snelling, R.R. 1988. Taxonomic notes on Nearctic species of *Camponotus*, subgenus *Myrmentoma* (Hymenoptera: Formicidae). In J.C. Trager, ed., Advances in myrmecology, pp. 55–78. Leiden: E.J. Brill.

Snelling, R.R. 2006. Review of *Carpenter ants of the United States and Canada*, by L.D. Hansen and J.H. Klotz. Notes from Underground 11(2), http://www.notesfromunderground.org/archive/issue11-2/index11-2.htm (accessed 28 January 2008).

Snelling, R.R., and C.D. George. 1979. The taxonomy, distribution and ecology of California desert ants (Hymenoptera: Formicidae). Report to California Desert Plan Program. Washington, D.C.: Bureau of Land Management, U.S. Department of the Interior.

Snodgrass, R.E. 1956. Anatomy of the honey bee. Ithaca: Cornell University Press.

Soeprono, A.M., and M.K. Rust. 2004a. The effect of horizontal transfer of barrier insecticides to control Argentine ants (Hymenoptera: Formicidae). J. Econ. Entomol. 97: 1675–1681.

Soeprono, A.M., and M.K. Rust. 2004b. The effect of delayed toxicity of chemical barriers to control Argentine ants (Hymenoptera: Formicidae). J. Econ. Entomol. 97: 2021–2028.

Stanley, M.C., and W.A. Robinson. 2007. Relative attractiveness of baits to *Paratrechina longicornis* (Hymenoptera: Formicidae). J. Econ. Entomol. 100: 509–516.

Stein, M.R., H.G. Thorvilson, and J.J. Johnson 1990. Seasonal changes in bait preferences by red imported fire ant, *Solenopsis invicta* (Hymenoptera: Formicidae). Fla. Entomol. 73: 117–123.

Steinbrink, H. 1974. *Iridomyrmex* im DDR-Küstenbezirk. Angew. Parasitol. 15: 34–36.

Steinbrink, H. 2000. Ameisen im Umfeld des Menschen in Mecklenburg-Vorpommern (MV). Pest Contr. News 24: 26–27.

Steiner, F.M., B.C. Schlick-Steiner, and A. Buschinger. 2003. First record of unicolonial polygyny in *Tetramorium* cf. *caespitum* (Hymenoptera: Formicidae). Insectes Soc. 50: 98–99.

Steiner, F.M., B.C. Schlick-Steiner, J.C. Trager, K. Moder, M. Sanetra, E. Christian, and C. Stauffer. 2006. *Tetramorium tsushimae,* a new invasive ant in North America. Biol. Invasions 8(2): 117–123.

Sterling, W.L. 1978. Fortuitous biological suppression of the boll weevil by the red imported fire ant. Environ. Entomol. 7: 564–568.

Storz, S.R., and W.R. Tschinkel. 2004. Distribution, spread, and ecological associations of the introduced ant *Pheidole obscurithorax* in the southeastern United States. J. Insect Sci. 4: 12.

Stringer, C.E. Jr., C.S. Lofgren, and F.J. Bartlett. 1964. Imported fire ant toxic bait studies: evaluation of toxicants. J. Econ. Entomol. 57: 941–945.

Stumper, R. 1950. Études myrmecologiques. VIII. Examen chimique de quelques nids de *Lasius fuliginosus* Latr. Arch. Inst. Grand-Ducal de Luxembourg (Sec. Sci. nat., phys., math.) 19: 243–250.

Suarez, A.V., N.D. Tsutsui, D.A. Holway, and T.J. Case. 1999. Behavioral and genetic differentiation between native and introduced populations of the Argentine ant. Biol. Invasions 1: 43–53.

Taber, S.W. 1998. The world of the harvester ant. College Station: Texas A&M University Press.

Taber, S.W. 2000. Fire ants. College Station: Texas A&M University Press.

Talbot, M. 1943. Responses of the ant *Prenolepis imparis* to temperature and humidity changes. Ecology 24: 345–352.

Taniguchi, G., T. Thompson, and B. Sipes. 2005. Control of the big-headed ant, *Pheidole megacephala* (Hymenoptera: Formicidae), in pineapple cultivation using Amdro in bait stations. Sociobiology 45: 1–7.

Tennant, L.E., and S.D. Porter. 1991. Comparison of diets of two fire ant species (Hymenoptera: Formicidae): solid and liquid components. J. Entomol. Sci. 26: 451–465.

Thompson, C.R. 1990. Ants that have pest status in the United States. In R.K. Vander Meer, K. Jaffe, and A. Cedeno, eds., Applied myrmecology: a world perspective, pp. 51–67. Boulder, Colo.: Westview Press,

Thorvilson, R., and B. Rudd. 2001. Are landscaping mulches repellent to red imported fire ants? Southwest. Entomol. 26: 195–203.

Trager, J.C. 1984. A revision of the genus *Paratrechina* (Hymenoptera: Formicidae) of the continental United States. Sociobiology 9: 49–162.

Trager, J.C. 1991. A revision of the fire ants, *Solenopsis geminata* group (Hymenoptera: Formicidae: Myrmicinae). J. N.Y. Entomol. Soc. 99: 141–198.

Tripp, J.M., D.R. Suiter, G.W. Bennett, J.H. Klotz, and B.L. Reid. 2000. Evaluation and control measures for black carpenter ant (Hymenoptera: Formicidae). J. Econ. Entomol. 93: 1493–1497.

Tschinkel, W.R. 1987. Seasonal life history and nest architecture of a winter-active ant, *Prenolepis imparis.* Insectes Soc. 34: 143–164.

Tschinkel, W.R. 2006. The fire ants. Cambridge, Mass.: Belknap Press of Harvard University Press.

Tschinkel, W.R., and D.F. Howard. 1980. A simple, non-toxic home remedy against fire ants. J. Ga. Entomol. Soc. 15: 102–105.

Tsutsui, N.D., and A.V. Suarez. 2003. The colony structure and population biology of invasive ants. Conserv. Biol. 17: 48–58.

Ulloa-Chacon, P., and D. Cherix. 1990. The little fire ant *Wasmannia auropunctata* (R.) (Hymenoptera: Formicidae). In R.K. Vander Meer, K. Jaffe, and A. Cedeno, eds., Applied myrmecology: a world perspective, pp. 281–289. Boulder, Colo.: Westview Press.

Ulloa-Chacon, P., and G.I. Jaramillo. 2003. Effects of boric acid, fipronil, hydramethylnon, and diflubenzuron baits on colonies of ghost ants (Hymenoptera: Formicidae). J. Econ. Entomol. 96: 856–862.

Umwelt- und Gesundheitsschutz Zürich. 2004. Relevanz der Pharaoameise in der Stadt Zürich und der Schweiz. Zürich, Switzerland: Agency of Environment and Healthcare.

U.S. Department of Agriculture. Agricultural Research Service. 2005. Areawide fire ant suppression, htttp://www.ars.usda.gov/fireant/preliminary.htm (accessed 1 February 2008).

Vail, K.M. 1997. Pharaoh ant control in large institutions: or spraying is not for pharaoh ants. Univ. Tenn. Agric. Ext. Serv. Publ. 623.

Vail, K.M., D. Bailey, and M. McGinnis. 2003. Perimeter spray and bait combo. Pest Contr. Tech. 31: 96–100.

Vail, K., L. Davis, D. Wojcik, P. Koehler, and D. Williams. 1994. Structure-invading ants of Florida. Coop. Ext. Serv. Bull. SP 164. Gainesville: University of Florida Institute of Food and Agricultural Sciences.

Vail, K.M., and D.F. Williams. 1994. Foraging of the pharaoh ant, *Monomorium pharaonis:* an exotic in the urban environment. In D.F. Williams, ed., Exotic ants: biology, impact, and control of introduced species, pp. 228–239. Boulder, Colo.: Westview Press.

Vail, K.M., D.F. Williams, and D.H. Oi. 1996. Perimeter treatments with two bait formulations of pyriproxfen for control of pharaoh ants (Hymenoptera: Formicidae). J. Econ. Entomol. 89: 1501–1507.

Valles, S.M., and P.G. Koehler. 2003. Insecticides used in the urban environment: mode of action. Florida Cooperative Extension Service, Entomology and Nematology Department, Publ. ENY282. Gainesville: University of Florida Institute of Food and Agricultural Sciences.

Valles, S.M., and R.M. Pereira. 2003. Hydramethylnon potentiation in *Solenopsis invicta* by infection with the microsporidium *Thelohania solenopsae*. Biol. Contr. 27: 9599.

Vander Meer, R.K., J.A. Seawright, and W.A. Banks. 1993. The use of repellents for area exclusion of pest ants. In K.B. Wildey and W.H. Robinson, eds., Proceedings of the First International Conference on Insect Pests in the Urban Environment, Cambridge, England, 30 June–3 July 1993, p. 494.

Vander Meer, R.K., D.F. Williams, and C.S. Lofgren: 1982. Degradation of the toxicant AC 217,300 in Amdro imported fire ant bait under field conditions. J. Agric. Food Chem. 30: 1045–1048.

Vargo, E.L., and S.D. Porter. 1989. Colony reproduction by budding in the polygyne form of *Solenopsis invicta* (Hymenoptera: Formicidae). Ann. Entomol. Soc. Am. 82: 307–313.

Vega, S.Y., and M.K. Rust. 2001. The Argentine ant—a significant invasive species in agriculture, urban and natural environments. Sociobiology 37: 3–25.

Vega, S.J., and M.K. Rust. 2003. Determining the foraging range and origin of resurgence after treatment of Argentine ant (Hymenoptera: Formicidae) in urban areas. J. Econ. Entomol. 96: 844–849.

Vinson, S.B. 1997. Invasion of the red imported fire ant (Hymenoptera: Formicidae): spread, biology, and impact. Am. Entomol. 43: 23–39.

Vinson, S.B., and L. Greenberg. 1986. The biology, physiology and ecology of imported fire ants. In S.B. Vinson, ed., Economic impact and control of social insects, pp. 193–226. New York: Praeger.

Vogt, J.T., J.T. Reed, and R.L. Brown. 2004. Temporal foraging activity of selected ant species in northern Mississippi during summer months. J. Entomol. Sci. 39: 444–451.

Vogt, J.T., J.T. Reed, and R.L. Brown. 2005. Timing bait applications for control of imported fire ants (Hymenoptera: Formicidae) in Mississippi: efficacy and effects on non-target ants. Int. J. Pest Manage. 51: 121–130.

Vogt, J.T., T.G. Shelton, M.E. Merchant, S.A. Russell, M.J. Tanley, and A.G. Appel. 2002. Efficacy of three citrus oil formulations against *Solenonsis invicta* Buren (Hyemnoptera: Formicidae), the red imported fire ant. J. Agric. Urban Entomol. 19: 159–171.

Vogt, J.T., D.A. Street, R.M. Pereira, and A.-M.A. Callcott. 2003. Mississippi areawide fire ant suppression program: unique aspects of working with black and hybrid imported fire ants. J. Agric. Urban Entomol. 20: 105–111.

Wagner, R.E. 1983. Effects of Amdro fire ant insecticide mound treatments on southern California ants, 1982. Insecticide & Acaricide Tests 8: 257.

Ward, P.S. 2005. A synoptic review of the ants of California (Hymenoptera: Formicidae). Zootaxa 936: 1–68.

Warner, J., and R.H. Scheffrahn. 2004a. Feeding preferences of white-footed ants, *Technomyrmex albipes* (Hymenoptera: Formicidae), to selected liquids. Sociobiology 44: 403–412.

Warner, J., and R.H. Scheffrahn. 2004b. Featured creatures: Caribbean crazy ant (proposed common name), *Paratrechina pubens* Forel (Insecta: Hymenoptera: Formicidae: Formicinae). Publ. EENY 284. Gainesville: University of Florida Institute of Food and Agricultural Sciences, http://creatures.ifas.ufl.edu/urban/ants/caribbean_crazy_ant.htm (accessed 28 January 2008).

Warner, J., and R.H. Scheffrahn. 2005. Laboratory evaluation of baits, residual insecticides, and an ultrasonic device for control of white-footed ants, *Technomyrmex albipes* (Hymenoptera: Formicidae). Sociobiology 45: 317–330.

Warner, J., R.H. Scheffrahn, and B. Cabrera. 2004. Featured creatures: white-footed ant. Gainesville: University of Florida Institute of Food and Agricultural Sciences, http://creatures .ifas.ufl.ufl.edu/urban/ants/white-footed_ant.htm (accessed 28 January 2008).

Weeks, R.D. Jr., L.T. Wilson, S.B. Vinson, and W.D. James. 2004. Flow of carbohydrates, lipids, and protein among colonies of polygyne red imported fire ants, *Solenopsis invicta* (Hymenoptera: Formicidae). Ann. Entomol. Soc. Am. 1135–110.

Wegner, G. 1991. The small honey ant. Pest Manage. 10: 28–29.

Weidner, H., and U. Sellenschlo. 2003. Vorratsschädlinge und Hausungeziefer. Heidelberg: Ed. Spektrum Akademischer Verlag.

Wetterer, J.K. 2005. Worldwide distribution and potential spread of the long-legged ant, *Anoplolepis gracilipes* (Hymenoptera: Formicidae). Sociobiology 45: 77–97.

Wetterer, J.K., S.E. Miller, D.E. Wheeler, C.A. Olson, D.A. Polhemus, M. Pitts, I.W. Ashton, A.G. Himler, M.M. Yospin, K.R. Helms, E.L. Harken, J. Gallager, C.E. Dunning, M. Nelson, J. Litsinger, A. Southern, and T.L. Burgess. 1999. Ecological dominance by *Paratrechina longicornis* (Hymenoptera: Formicidae), an invasive tramp ant, in Biosphere 2. Fla. Entomol. 82: 381–388.

Wetterer, J.K., and S.D. Porter. 2003. The little fire ant, *Wasmannia auropunctata:* distribution, impact and control. Sociobiology 42: 1–41.

Wheeler, G.C., and J.N. Wheeler. 1963. The ants of North Dakota. Grand Forks: University of North Dakota Press.

Wheeler, G.C., and J.N. Wheeler. 1973. Ants of Deep Canyon, Colorado Desert, California. Philip L. Boyd Deep Canyon Desert Research Center, University of California, Riverside.

Wheeler, G.C., and J.N. Wheeler. 1986. The ants of Nevada. Los Angeles: Natural History Museum of Los Angeles County.

Wheeler, J., and S.W. Rissing. 1975. Natural history of *Veromessor pergandei*. II. Behavior. Pan-Pac. Entomol. 51: 303–314.

Wheeler, W.M. 1905. The North American ants of the genus *Liometopum*. Bull. Am. Mus. Nat. Hist. 21: 325–333.

Whitford, M. 2006. Up to the challenge. Pest Contr. 74(9): S11—S15.

Wild, A.L. 2004. Taxonomy and distribution of the Argentine ant, *Linepithema humile* (Hymenoptera: Formicidae). Ann. Entomol. Soc. Am. 97: 1204–1215.

Wildermuth, V.L., and E.G. Davis. 1931. The red harvester ant and how to subdue it. USDA Farmer's Bull. 1668. Washington, D.C.: U.S. Department of Agriculture.

Williams, D.F., W.A. Banks, and C.S. Lofgren. 1997. Control of *Solenopsis invicta* (Hymenoptera: Formicidae) with teflubenzuron. Fla. Entomol. 80: 84–91.

Williams, D.F., H.L. Collins, and D.H. Oi. 2001. The red imported fire ant (Hymenoptera: Formicidae): a historical perspective of treatment programs and the development of chemical baits for control. Am. Entomol. 47: 146–159.

Williams, D.F., and R.D. deShazo. 2004. Biological control of fire ants: an update on new techniques. Ann. Allergy, Asthma, and Immunol. 93: 15–22.

Willams, D.F., D.H. Oi, S.D. Porter, R.M. Pereira, and J.A. Briano. 2003. Biological control of imported fire ants. Am. Entomol. 49: 150–163.

Williams, D.F., and K.M. Vail. 1993. Pharaoh ant (Hymenoptera: Formicidae); fenoxycarb baits affect colony development. J. Econ. Entomol. 86: 1136–1143.

Williams, D.F., and P.M. Whelan. 1992. Bait attraction of the introduced pest ant, *Wasmannia auropunctata* (Hymenoptera: Formicidae), in the Galápagos Islands. J. Entomol. Sci. 27: 29–34.

Wilson, E.O. 1955. A monographic revision of the ant genus *Lasius*. Bull. Mus. Comp. Zool. 113(1): 1–199.

Wilson, E.O. 1971. The insect societies. Cambridge, Mass.: Belknap Press of Harvard University Press.

Wilson, E.O. 1976. Which are the most prevalent ant genera? Studia Entomol. 19: 187–200.

Wilson, E.O. 2003. *Pheidole* in the New World. Cambridge, Mass.: Harvard University Press.

Wilson, E.O., and R.W. Taylor. 1967. The ants of Polynesia (Hymenoptera: Formicidae). Pac. Insects Monogr. 14. Honolulu: Bernice P. Bishop Museum, Entomology Department.

Wojcik, D.P., C.R. Allen, R.J. Brenner, E.A. Forys, D.P. Jouvenaz, and R.S. Lutz. 2001. Red imported fire ants: impact on biodiversity. Am. Entomol. 47: 16–23.

Yamauchi, K., T. Furukawa, K. Kinomura, H. Takamine, and K. Tsuji. 1991. Secondary polygyny by inbred wingless sexuals in the dolichoderine ant *Technomyrmex albipes*. Behav. Ecol. Sociobiol. 29: 313–319.

Young, J., and D.E. Howell. 1964. Ants of Oklahoma. Okla. Agric. Exp. Sta. Misc. Publ. MP-71.

Zarzuela, M.F., A.E. de C. Campos-Farinha, and M.P. Pecanha. 2005. Evaluation of urban ants (Hymenoptera: Formicidae) as carriers of pathogens in residential and industrial environments: I. Bacteria. Sociobiology 45: 9–14.

Zhao, X., T. Ikeda, V.L. Salgado, J.Z. Yeh, and T. Narahashi. 2005. Block of two subtypes of sodium channels in cockroach neurons by indoxacarb insecticides. NeuroToxicology 26: 455–465.

Zungoli, P., E. Paysen, E. Benson, and J. Nauman. 2005. Colony and habitat characteristics of *Pachycondyla chinensis* (Hymenoptera: Formicidae). In C. Lee and W.H. Robinson, eds., Proceedings of the Fifth International Conference on Urban Pests. Penang, Malaysia: Perniagaan Ph'ng@P&Y Design Network.

# Index